BLUEPRINT READING AND SKETCHING FOR CARPENTERS

RESIDENTIAL

BLUEPRINT READING AND SKETCHING FOR CARPENTERS

RESIDENTIAL

LEO McDONNELL
JOHN E. BALL

18 19 HAM/HAM 0 9 8 7 6 5 4

LIBRARY OF CONGRESS CATALOG CARD NUMBER: 80-66027
ISBN: 0-8273-1354-3

Printed in the United States of America
Published simultaneously in Canada
by Nelson Canada,
A Division of International Thomson Limited

NOTICE TO THE READER

DELMAR PUBLISHERS INC. • ALBANY, NEW YORK

PREFACE

Blueprints, trade sketches, and specifications are the instructions followed by the mason, carpenter, plumber, electrician, and others on a construction project. The contractor, subcontractor, tradesworker, and apprentice receive their information from a common set of plans and specifications. All, therefore, share a common need for the ability to interpret trade blueprints and to make on-the-job sketches. For years, BLUE-PRINT READING AND SKETCHING FOR CARPENTERS — RESIDENTIAL has fulfilled the need for a tool to accomplish these teaching tasks.

The New Edition

This latest edition of the text has many new features, including:

- a new unit on symbols and abbreviations
- an expanded unit on specifications
- a discussion of slab-on-grade, crawl space, and basements
- up-to-date symbol charts
- new units on heating and air-conditioning plans, plumbing plans, and electrical plans

The book is designed to serve both as a text and a study guide. When the course has been completed, it may serve as a reference book in which the student's work and solutions to all problem material are recorded. The content is readily adaptable to trade preparatory courses and apprentice training and for upgrading training through occupational extension work and home study.

The Individual Unit

Objectives at the beginning of each unit state what the student is expected to accomplish before advancing to new material. Illustrations are numerous. Technical information is presented in simple trade language at an understandable reading level. At the conclusion of each unit, there is an assignment which provides problem material in blueprint interpretation and, in most cases, sketching problems so that the student may further apply principles learned through study of the unit. Each sketching assignment includes a ruled form on which the student can work.

The units are arranged in logical sequence for advancement from one unit to the next. When the student group has similar backgrounds and study goals, the instructor should follow the sequence of units as they are presented. However, when the group has widely diversified backgrounds, the instructor may find it necessary to use selected units. Each unit is complete within itself and, if necessary, their order may be rearranged to conform to local conditions.

The Appendix and House Plans

Included in the Appendix is a complete set of specifications for the carpentry and masonry work involved in the construction of a house. Seven blueprints of this house appear in the packet at the back of the book. These drawings are used in conjunction with the instructional material.

Instructor's Guide

An Instructor's Guide to this text is available. The Guide includes the answers to all assignment questions in the units and in the comprehensive review. A suggested time chart gives the approximate amount of time required for the average student to complete each unit. There are also visual aids which the instructor may reproduce for classroom use.

Related Delmar Titles

The following are other construction-related texts which may prove helpful to the student.

BASIC CONSTRUCTION BLUEPRINT READING
BLUEPRINT READING FOR PLUMBERS
SHEET METAL BLUEPRINT READING FOR THE BUILDING TRADES
BLUEPRINT READING FOR COMMERCIAL CONSTRUCTION
WORKING DRAWINGS FOR COMMERCIAL CONSTRUCTION
FRAMING, SHEATHING & INSULATION
EXTERIOR AND INTERIOR TRIM
SIMPLIFIED STAIR LAYOUT
INTRODUCTION TO CONSTRUCTION
MATHEMATICS FOR CARPENTERS
PRACTICAL PROBLEMS IN MATHEMATICS FOR CARPENTERS
ESTIMATING FOR RESIDENTIAL CONSTRUCTION
MASONRY SKILLS
ADVANCED MASONRY SKILLS
CONCRETE FORM CONSTRUCTION
BASIC PLUMBING
ADVANCED PLUMBING
CONSTRUCTION ESTIMATING

TABLE OF CONTENTS

Unit

UNIT 1 SYMBOLS AND ABBREVIATIONS

OBJECTIVES

- Explain symbols used in plumbing, heating, electrical, and carpentry work.
- List abbreviations commonly used in the construction industry.

Architects use a set of working drawings (blueprints) and specifications to convey their instructions to skilled workers who erect the structure. Construction features, such as the size and location of the structure and quantities of necessary materials, are shown at a glance in the drawings. This simplifies understanding of the blueprints and specifications provided by the architect.

It is the carpenters' responsibility to identify the plans and specifications. Not only must they understand those lines and symbols which pertain to their work, but also those which apply to the mason, electrician, plumber, and other workers who all use a common set of plans.

Certain features shown on the architectural drawings are indicated by use of a symbol. The symbols are standardized and carry the same meaning from one set of drawings to another.

Abbreviations are often used in the plans to save time. However, abbreviations are often confusing because of their inconsistencies. When they are used, they are spelled in capital letters with no period at the end of the word.

Figures 1-1 through 1-5 illustrate symbols and abbreviations that are commonly used in the construction industry.

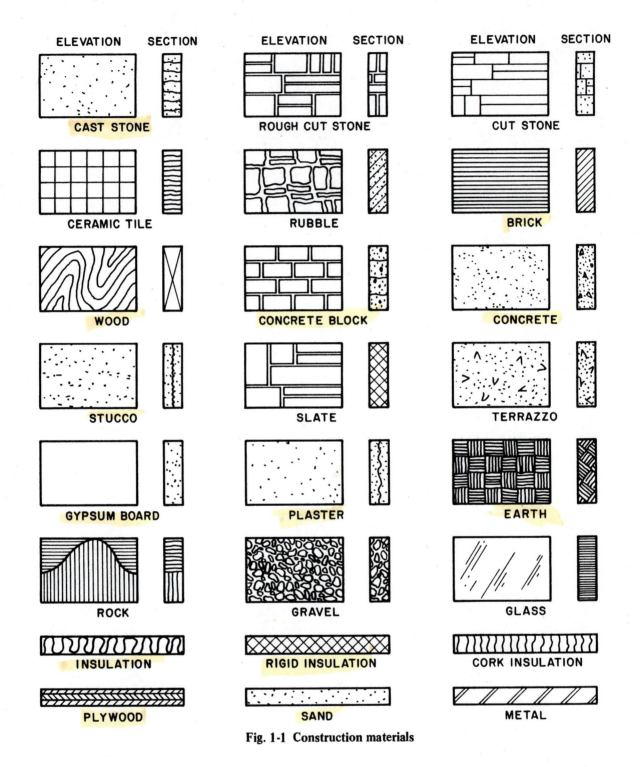

Fig. 1-1 Construction materials

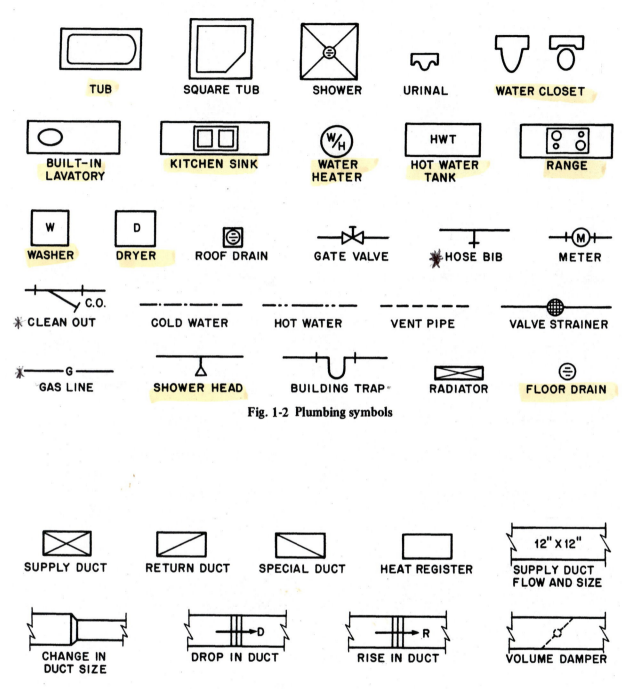

Fig. 1-2 Plumbing symbols

TUB

SQUARE TUB

SHOWER

URINAL

WATER CLOSET

BUILT-IN LAVATORY

KITCHEN SINK

WATER HEATER

HOT WATER TANK

RANGE

WASHER

DRYER

ROOF DRAIN

GATE VALVE

HOSE BIB

METER

CLEAN OUT

COLD WATER

HOT WATER

VENT PIPE

VALVE STRAINER

GAS LINE

SHOWER HEAD

BUILDING TRAP

RADIATOR

FLOOR DRAIN

SUPPLY DUCT

RETURN DUCT

SPECIAL DUCT

HEAT REGISTER

12" X 12"

SUPPLY DUCT FLOW AND SIZE

CHANGE IN DUCT SIZE

DROP IN DUCT

RISE IN DUCT

VOLUME DAMPER

Fig. 1-3 Air ducts

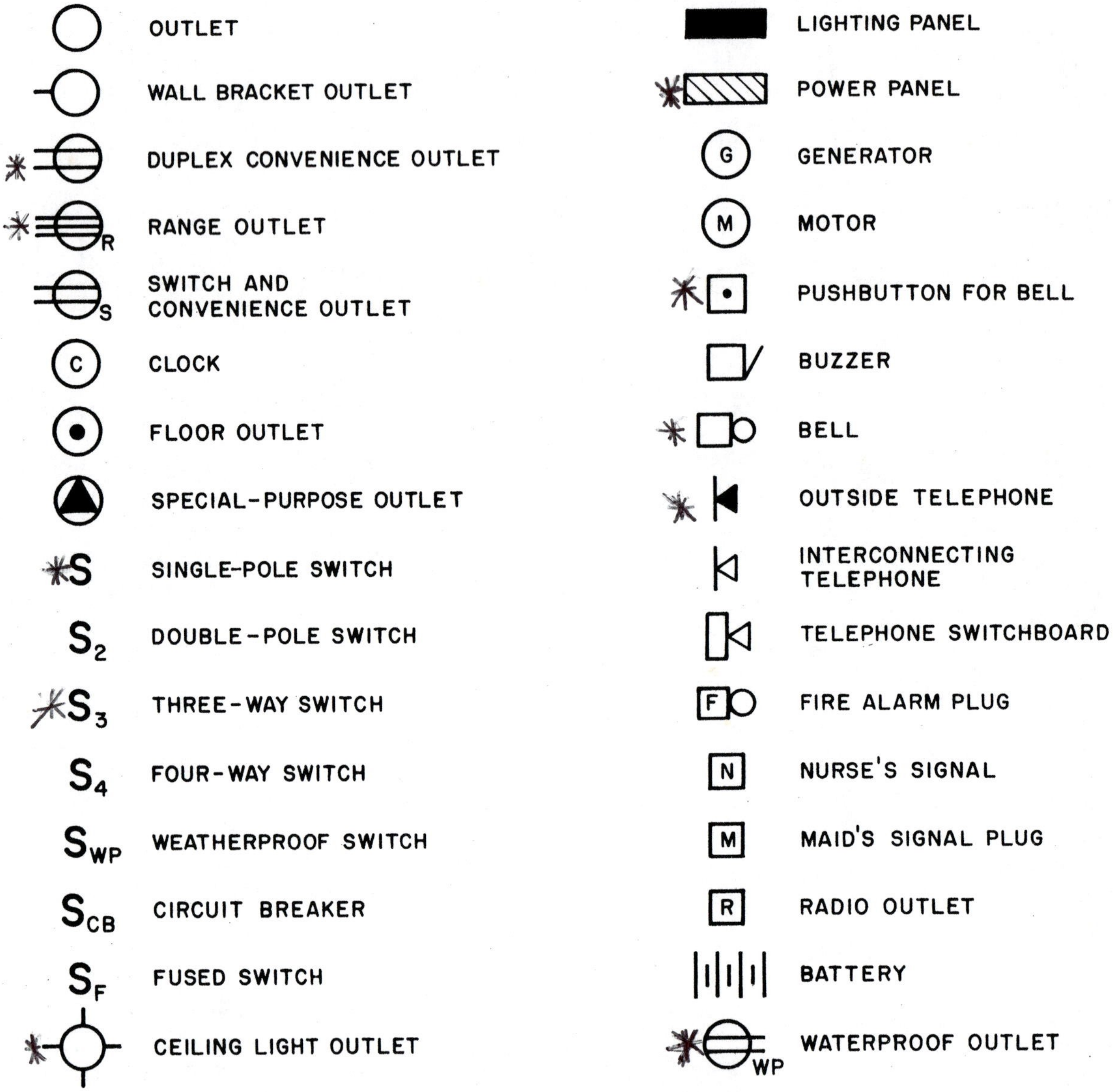

◯	OUTLET
─◯	WALL BRACKET OUTLET
✳═⊖	DUPLEX CONVENIENCE OUTLET
✳═⊜$_R$	RANGE OUTLET
⊖$_S$	SWITCH AND CONVENIENCE OUTLET
ⓒ	CLOCK
⊙	FLOOR OUTLET
⬤	SPECIAL-PURPOSE OUTLET
✳S	SINGLE-POLE SWITCH
S$_2$	DOUBLE-POLE SWITCH
✳S$_3$	THREE-WAY SWITCH
S$_4$	FOUR-WAY SWITCH
S$_{WP}$	WEATHERPROOF SWITCH
S$_{CB}$	CIRCUIT BREAKER
S$_F$	FUSED SWITCH
✳-◯-	CEILING LIGHT OUTLET

▅	LIGHTING PANEL
✳▨	POWER PANEL
Ⓖ	GENERATOR
Ⓜ	MOTOR
✳▣	PUSHBUTTON FOR BELL
◻╱	BUZZER
✳◻	BELL
✳◀	OUTSIDE TELEPHONE
◁	INTERCONNECTING TELEPHONE
◁	TELEPHONE SWITCHBOARD
F◯	FIRE ALARM PLUG
N	NURSE'S SIGNAL
M	MAID'S SIGNAL PLUG
R	RADIO OUTLET
∣∣∣∣	BATTERY
✳⊖$_{WP}$	WATERPROOF OUTLET

Fig. 1-4 Electrical symbols

access door	AD	dowel	DWL
access panel	AP	down	DN
acoustic	ACST	downspout	DS
actual	ACT	drawing	DWG
air conditioning	AC	each	EA
alternating current	AC	east	E
approximate	APPROX	electric	ELEC
architect	ARCH	elevation	EL
architectural	ARCH	enclose	ENCL
article	ART	engineer	ENGR
asphalt	ASPH	entrance	ENT
associate	ASSOC	equipment	EQUIP
at	@	estimate	EST
average	AVG	exterior	EXT
beam	BM	fabricate	FAB
bedroom	BR	feet	FT
bench mark	BM	feet board measure	FBM
better	BTR	finish	FIN
between	BET	finished floor	FL
beveled	BEV	fire extinguisher	F EXT
block	BLK	fireproof	FPRF
board feet	BD FT	flange	FLG
bottom	BOT	flashing	FL
British thermal unit	BTU	floor	FL
building	BLDG	floor drain	FD
built-in	BLT-IN	fluorescent	FLUOR
cabinet	CAB	foot	FT
caulking	CLKG	footing	FTG
carpenter	CARP	foundation	FDN
ceiling	CLG	frame	FR
cement	CEM	full size	FS
change	CHG	gallon	GAL
closet	CLO	galvanized	GALV
cold water	CW	general contractor	GEN CONT
column	COL	glass	GL
common	COM	government	GOVT
company	CO	grade	GR
concrete	CONC	grade line	GL
concrete floor	CONC FL	gypsum	GYP
construction	CONST	hall	H
cover plate	COV PL	half-round	H RD
cubic foot	CU FT	hardwood	HDWD
dampproofing	DP	head	HD
degree	DEG	heater	HTR
department	DEPT	height	HGT
diameter	DIA	hollow metal	HM
distance	DIST	hollow metal door	HMD
ditto	DO	horizon	H
division	DIV	horizontal	HOR
double hung window	DHW	hose bibb	HB

Fig. 1-5 Abbreviations

hospital	HOSP	platform	PLAT
hot water	HW	plumbing	PLMB
house	HSE	point	PT
inch	IN	polished plate glass	PPGL
include	INCL	pound	LB
incorporated	INC	pounds per cubic foot	PCF
information	INFO	pounds per square foot	PSF
inside diameter	ID	precast	PRCST
inside pipe size	IPS	property	PROP
insulation	INS	proposed	PROP
interior	INT	quantity	QTY
janitor's closet	J CL	radiator	RAD
joint	JT	radius	RAD
kip	K	random	RDM
kitchen	KIT	receptacle	RECP
knocked down	KD	register	REG
laboratory	L	reinforcing bar	REBAR
left hand	LH	required	REQD
length	LG	revision	REV
linear feet	LIN FT	road	RD
live load	LL	roofing	RFG
lumber	LBR	room	RM
machine	MACH	rough	RGH
manhole	MH	round	RD
manufacture	MFR	scale	SC
masonry opening	MO	schedule	SCH
material	MATL	screw	SCR
mechanical	MECH	section	SECT
medium	MED	select	SEL
membrane	MEMB	sewer	SEW
metal	MET	sheathing	SHTHG
minimum	MIN	sheet	SH
miscellaneous	MISC	shower	SH
molding	MLDG	siding	SDG
not in contract	NIC	sink	SK
nominal	NOM	sink, service	SS
north	N	sleeve	SLV
number	NO	service sink	SS
office	OFF	soil pipe	SP
on center	OC	south	S
one-thousand feet board measure	MBM	speaker	SPKR
opening	OPNG	specifications	SPEC
out to out	O to O	square	SQ
pair	PR	square foot	SQ FT
panel	PNL	square inch	SQ IN
penny	d	stained	STN
per	/	staggered	STAG
piece	PC	stained-waxed	SW
plaster	PLAS	stainless steel	SST
plastic	PLSTC	stairs	ST

Fig. 1-5 Abbreviations (Cont'd.)

⌐ standard	⌐STD	unfinished	UNFIN
⌐ steel	⌐STL	vaporproof	VAP PRF
⌐ street	⌐ST	variable	VAR
⌐structural	⌐STR	ventilation	VENT
substitute	SUB	vertical	VERT
superintendent	SUPT	vestibule	VEST
supply	SUP	vitreous	VIT
⌐ surface	⌐SUR	volume	VOL
⌐ surfaced four sides	⌐ S4S	⌐ water closet	⌐ WC
⌐ suspended ceiling	⌐ SUSP CEIL	⌐ weatherproof	⌐ WP
⌐ switch	⌐ SW	weight	WT
switchboard	SWBD	⌐ west	⌐ W
symbol	SYM	wide flange (steel)	WF
telephone	TEL	width	W
temperature	TEMP	⌐ window	⌐ WDW
template	TEMP	wire glass	W GL
thermostat	THERMO	wood door	WD
thousand	M	wood frame	WF
⌐ thousand pounds	⌐ KIP	⌐ yard	⌐ YD
⌐tongue and groove	⌐T&G	yellow pine	YP
⌐ typical	⌐ TYP	zinc	ZN

Fig. 1-5 Abbreviations (Cont'd.)

ASSIGNMENT

Identify the following symbols and abbreviations.

1. _____

2. _____

3. _____

4. _____

5. _____

6. _____

7. _____

8. _____

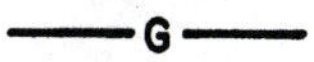

9. _____

10. _____

11. _____

12. _____

13. _____

14. _____

15. _____

16. _____

17. _____

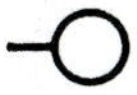

18. _____

19. _____

20. AC _____

21. DEPT _____

22. DO _____

23. EA   _____   30. MIN   _____

24. FL   _____   31. MLDG   _____

25. GAL   _____   32. RD   _____

26. GYP   _____   33. SC   _____

27. HB   _____   34. T&G   _____

28. JT   _____   35. S4S   _____

29. LH   _____

UNIT 2 SPECIFICATIONS

OBJECTIVES

- Explain information which is found in specifications.
- List three purposes of specifications.

Most working drawings become quite complex because of the great amount of data which must be included in them. To avoid additional complexity, it is standard practice to include detailed written instructions with each set of plans. These are called the *specifications.*

Specifications are the written interpretation of the drawings. They define the various jobs to be accomplished in the design and construction of the building. They are an important part of the contract and, in almost all cases, supersede the drawings. This is because once the contract is signed and executed, the specifications cannot be changed. However, corrections can be made on the drawings without breaking the contract.

The drawings and specifications are read simultaneously. The drawings show sizes of features and the layout of the project. Specifications explain the drawings and specify the materials to be used, the methods of application, and the quality and extent of the work. In this respect, they are a great help to the contractor in estimating cost, ordering required materials, and explaining items that are difficult to show on the drawings. Inspections, rejections, and approvals are definitely stated so that the contractor knows exactly what is called for regarding cost and labor.

Specifications are prepared by the architect. They cover the entire project, with the information generally grouped by trades. Typical specifications have a section describing the general clauses and conditions which apply to all trades. These are followed by descriptions of the requirements for specific trades, such as excavating, masonry, carpentry, plumbing, heating, electrical work, and painting.

Besides being grouped by trades, specifications are usually written in the order in which the jobs involved in the construction are to be performed. They should also be prepared so that the user can quickly and accurately find any information that might be needed.

Purpose of Specifications

Specifications serve several purposes.

- They help avoid dispute between the builder and the customer or between the builder and the architect.
- They eliminate conflicting opinions regarding the grade of material to be used or the quality of work required. If the specifications and the working drawings differ on a given item, the specifications are followed.
- They help the contractor estimate material and labor.

9

FHA Form 2005
VA Form 26-1852
Rev. 2/75

U. S. DEPARTMENT OF HOUSING AND URBAN DEVELOPMENT
FEDERAL HOUSING ADMINISTRATION

For accurate register of carbon copies, form
may be separated along above fold. Staple
completed sheets together in original order.

Form Approved
OMB No. 63—R0055

☐ Proposed Construction

DESCRIPTION OF MATERIALS   No. _____
(To be inserted by FHA or VA)

☐ Under Construction

Property address _____ City _____ State _____

Mortgagor or Sponsor _____ _____
(Name)   (Address)

Contractor or Builder _____ _____
(Name)   (Address)

INSTRUCTIONS

1. For additional information on how this form is to be submitted, number of copies, etc., see the instructions applicable to the FHA Application for Mortgage Insurance or VA Request for Determination of Reasonable Value, as the case may be.

2. Describe all materials and equipment to be used, whether or not shown on the drawings, by marking an X in each appropriate check-box and entering the information called for in each space. If space is inadequate, enter "See misc." and describe under item 27 or on an attached sheet. THE USE OF PAINT CONTAINING MORE THAN ONE HALF OF ONE PERCENT LEAD BY WEIGHT IS PROHIBITED.

3. Work not specifically described or shown will not be considered unless

required, then the minimum acceptable will be assumed. Work exceeding minimum requirements cannot be considered unless specifically described.

4. Include no alternates, "or equal" phrases, or contradictory items. (Consideration of a request for acceptance of substitute materials or equipment is not thereby precluded.)

5. Include signatures required at the end of this form.

6. The construction shall be completed in compliance with the related drawings and specifications, as amended during processing. The specifications include this Description of Materials and the applicable Minimum Property Standards.

1. EXCAVATION:
Bearing soil, type _____

2. FOUNDATIONS:
Footings: concrete mix _____ ; strength psi _____ Reinforcing _____
Foundation wall: material _____ Reinforcing _____
Interior foundation wall: material _____ Party foundation wall _____
Columns: material and sizes _____ Piers: material and reinforcing _____
Girders: material and sizes _____ Sills: material _____
Basement entrance areaway _____ Window areaways _____
Waterproofing _____ Footing drains _____
Termite protection _____
Basementless space: ground cover _____ ; insulation _____ ; foundation vents _____
Special foundations _____
Additional information: _____

3. CHIMNEYS:
Material _____ Prefabricated (make and size) _____
Flue lining: material _____ Heater flue size _____ Fireplace flue size _____
Vents (material and size): gas or oil heater _____ ; water heater _____
Additional information: _____

4. FIREPLACES:
Type: ☐ solid fuel; ☐ gas-burning; ☐ circulator (make and size) _____ Ash dump and clean-out _____
Fireplace: facing _____ ; lining _____ ; hearth _____ ; mantel _____
Additional information: _____

5. EXTERIOR WALLS:
Wood frame: wood grade, and species _____ ☐ Corner bracing. Building paper or felt _____
Sheathing _____ ; thickness _____ ; width _____ ; ☐ solid; ☐ spaced _____ " o. c.; ☐ diagonal; _____
Siding _____ ; grade _____ ; type _____ ; size _____ ; exposure _____ "; fastening _____
Shingles _____ ; grade _____ ; type _____ ; size _____ ; exposure _____ "; fastening _____
Stucco _____ ; thickness _____ "; Lath _____ ; weight _____ lb.
Masonry veneer _____ Sills _____ Lintels _____ Base flashing _____
Masonry: ☐ solid ☐ faced ☐ stuccoed; total wall thickness _____ "; facing thickness _____ "; facing material _____
Backup material _____ ; thickness _____ "; bonding _____
Door sills _____ Window sills _____ Lintels _____ Base flashing _____
Interior surfaces: dampproofing, _____ coats of _____ ; furring _____
Additional information: _____
Exterior painting: material _____ ; number of coats _____
Gable wall construction: ☐ same as main walls; ☐ other construction _____

6. FLOOR FRAMING:
Joists: wood, grade, and species _____ ; other _____ ; bridging _____ ; anchors _____
Concrete slab: ☐ basement floor; ☐ first floor; ☐ ground supported; ☐ self-supporting; mix _____ ; thickness _____ ";
reinforcing _____ ; insulation _____ ; membrane _____
Fill under slab: material _____ ; thickness _____ ". Additional information: _____

7. SUBFLOORING: (Describe underflooring for special floors under item 21.)
Material: grade and species _____ ; size _____ ; type _____
Laid: ☐ first floor; ☐ second floor; ☐ attic _____ sq. ft.; ☐ diagonal; ☐ right angles. Additional information: _____

8. FINISH FLOORING: (Wood only. Describe other finish flooring under item 21.)

LOCATION	ROOMS	GRADE	SPECIES	THICKNESS	WIDTH	BLDG. PAPER	FINISH
First floor							
Second floor							
Attic floor _____ sq. ft.							
Additional information:							

Fig. 2-1 FHA Form 2005

DESCRIPTION OF MATERIALS

9. PARTITION FRAMING:
Studs: wood, grade, and species _____ size and spacing _____ Other _____
Additional information: _____

10. CEILING FRAMING:
Joists: wood, grade, and species _____ Other _____ Bridging _____
Additional information: _____

11. ROOF FRAMING:
Rafters: wood, grade, and species _____ Roof trusses (see detail): grade and species _____
Additional information: _____

12. ROOFING:
Sheathing: wood, grade, and species _____ ; ☐ solid; ☐ spaced _____ " o.c.
Roofing _____ ; grade _____ ; size _____ ; type _____
Underlay _____ ; weight or thickness _____ ; size _____ ; fastening _____
Built-up roofing _____ ; number of plies _____ ; surfacing material _____
Flashing: material _____ ; gage or weight _____ ; ☐ gravel stops; ☐ snow guards
Additional information: _____

13. GUTTERS AND DOWNSPOUTS:
Gutters: material _____ ; gage or weight _____ ; size _____ ; shape _____
Downspouts: material _____ ; gage or weight _____ ; size _____ ; shape _____ ; number _____
Downspouts connected to: ☐ Storm sewer; ☐ sanitary sewer; ☐ dry-well. ☐ Splash blocks: material and size _____
Additional information: _____

14. LATH AND PLASTER
Lath ☐ walls, ☐ ceilings: material _____ ; weight or thickness _____ Plaster: coats _____ ; finish _____
Dry-wall ☐ walls, ☐ ceilings: material _____ ; thickness _____ ; finish _____ ;
Joint treatment _____

15. DECORATING: *(Paint, wallpaper, etc.)*

Rooms	Wall Finish Material and Application	Ceiling Finish Material and Application
Kitchen _____		
Bath _____		
Other _____		

Additional information: _____

16. INTERIOR DOORS AND TRIM:
Doors: type _____ ; material _____ ; thickness _____
Door trim: type _____ ; material _____ Base: type _____ ; material _____ ; size _____
Finish: doors _____ ; trim _____
Other trim *(item, type and location)* _____
Additional information: _____

17. WINDOWS:
Windows: type _____ ; make _____ ; material _____ ; sash thickness _____
Glass: grade _____ ; ☐ sash weights; ☐ balances, type _____ ; head flashing _____
Trim: type _____ ; material _____ Paint _____ ; number coats _____
Weatherstripping: type _____ ; material _____ Storm sash, number _____
Screens: ☐ full; ☐ half; type _____ ; number _____ ; screen cloth material _____
Basement windows: type _____ ; material _____ ; screens, number _____ ; Storm sash, number _____
Special windows _____
Additional information: _____

18. ENTRANCES AND EXTERIOR DETAIL:
Main entrance door: material _____ ; width _____ ; thickness _____ ". Frame: material _____ ; thickness _____ "
Other entrance doors: material _____ ; width _____ ; thickness _____ ". Frame: material _____ ; thickness _____ "
Head flashing _____ Weatherstripping: type _____ ; saddles _____
Screen doors: thickness _____ "; number _____ ; screen cloth material _____ Storm doors: thickness _____ "; number _____
Combination storm and screen doors: thickness _____ "; number _____ ; screen cloth material _____
Shutters: ☐ hinged; ☐ fixed. Railings _____ , Attic louvers _____
Exterior millwork: grade and species _____ Paint _____ ; number coats _____
Additional information: _____

19. CABINETS AND INTERIOR DETAIL:
Kitchen cabinets, wall units: material _____ ; lineal feet of shelves _____ ; shelf width _____
Base units: material _____ ; counter top _____ ; edging _____
Back and end splash _____ Finish of cabinets _____ ; number coats _____
Medicine cabinets: make _____ ; model _____
Other cabinets and built-in furniture _____
Additional information: _____

20. STAIRS:

Stair	Treads		Risers		Strings		Handrail		Balusters	
	Material	Thickness	Material	Thickness	Material	Size	Material	Size	Material	Size
Basement _____										
Main _____										
Attic _____										

Disappearing: make and model number _____
Additional information: _____

2

Fig. 2-1 FHA Form 2005 (Cont'd.)

21. SPECIAL FLOORS AND WAINSCOT:

	LOCATION	MATERIAL, COLOR, BORDER, SIZES, GAGE, ETC.	THRESHOLD MATERIAL	WALL BASE MATERIAL	UNDERFLOOR MATERIAL
FLOORS	Kitchen				
	Bath				

	LOCATION	MATERIAL, COLOR, BORDER, CAP. SIZES, GAGE, ETC.	HEIGHT	HEIGHT OVER TUB	HEIGHT IN SHOWERS (FROM FLOOR)
WAINSCOT	Bath				

Bathroom accessories: ☐ Recessed; material _____ ; number _____ ; ☐ Attached; material _____ ; number _____
Additional information: _____

22. PLUMBING:

FIXTURE	NUMBER	LOCATION	MAKE	MFR'S FIXTURE IDENTIFICATION NO.	SIZE	COLOR
Sink						
Lavatory						
Water closet						
Bathtub						
Shower over tub △						
Stall shower △						
Laundry trays						

△☐ Curtain rod   △☐ Door   ☐ Shower pan: material _____
Water supply: ☐ public; ☐ community system; ☐ individual (private) system. ★
Sewage disposal: ☐ public; ☐ community system; ☐ individual (private) system. ★
★ Show and describe individual system in complete detail in separate drawings and specifications according to requirements.
House drain (inside): ☐ cast iron; ☐ tile; ☐ other _____   House sewer (outside): ☐ cast iron; ☐ tile; ☐ other _____
Water piping: ☐ galvanized steel; ☐ copper tubing; ☐ other _____   Sill cocks, number _____
Domestic water heater: type _____ ; make and model _____ ; heating capacity
_____ gph. 100° rise. Storage tank: material _____ ; capacity _____ gallons.
Gas service: ☐ utility company; ☐ liq. pet. gas; ☐ other _____   Gas piping: ☐ cooking; ☐ house heating.
Footing drains connected to: ☐ storm sewer; ☐ sanitary sewer; ☐ dry well. Sump pump; make and model _____
_____ ; capacity _____ ; discharges into _____

23. HEATING:

☐ Hot water. ☐ Steam. ☐ Vapor. ☐ One-pipe system. ☐ Two-pipe system.
   ☐ Radiators. ☐ Convectors. ☐ Baseboard radiation. Make and model _____
   Radiant panel: ☐ floor; ☐ wall; ☐ ceiling. Panel coil: material _____
   ☐ Circulator. ☐ Return pump. Make and model _____ ; capacity _____ gpm.
   Boiler: make and model _____   Output _____ Btuh.; net rating _____ Btuh.
Additional information: _____
Warm air: ☐ Gravity. ☐ Forced. Type of system _____
   Duct material: supply _____ ; return _____   Insulation _____ , thickness _____ ☐ Outside air intake.
   Furnace: make and model _____   Input _____ Btuh.; output _____ Btuh.
   Additional information: _____
☐ Space heater; ☐ floor furnace; ☐ wall heater. Input _____ Btuh.; output _____ Btuh.; number units _____
   Make, model _____   Additional information: _____
Controls: make and types _____
Additional information: _____
Fuel: ☐ Coal; ☐ oil; ☐ gas; ☐ liq. pet. gas; ☐ electric; ☐ other _____ ; storage capacity _____
   Additional information: _____
Firing equipment furnished separately: ☐ Gas burner, conversion type. ☐ Stoker: hopper feed ☐; bin feed ☐
   Oil burner: ☐ pressure atomizing; ☐ vaporizing _____
   Make and model _____   Control _____
   Additional information: _____
Electric heating system: type _____   Input _____ watts; @ _____ volts; output _____ Btuh.
   Additional information: _____
Ventilating equipment: attic fan, make and model _____ ; capacity _____ cfm.
   kitchen exhaust fan. make and model _____
Other heating, ventilating, or cooling equipment _____

24. ELECTRIC WIRING:

Service: ☐ overhead; ☐ underground. Panel: ☐ fuse box; ☐ circuit-breaker; make _____ AMP's _____ No. circuits _____
Wiring: ☐ conduit; ☐ armored cable; ☐ nonmetallic cable; ☐ knob and tube; ☐ other _____
Special outlets: ☐ range; ☐ water heater; ☐ other _____
☐ Doorbell. ☐ Chimes. Push-button locations _____   Additional information: _____

25. LIGHTING FIXTURES:

Total number of fixtures _____   Total allowance for fixtures, typical installation, $ _____
Nontypical installation _____
Additional information: _____

DESCRIPTION OF MATERIALS

Fig. 2-1 FHA Form 2005 (Cont'd.)

DESCRIPTION OF MATERIALS

26. INSULATION:

Location	Thickness	Material, Type, and Method of Installation	Vapor Barrier
Roof			
Ceiling			
Wall			
Floor			

27. MISCELLANEOUS: (Describe any main dwelling materials, equipment, or construction items not shown elsewhere; or use to provide additional information where the space provided was inadequate. Always reference by item number to correspond to numbering used on this form.) _____

HARDWARE: (make, material, and finish.) _____

SPECIAL EQUIPMENT: (State material or make, model and quantity. Include only equipment and appliances which are acceptable by local law, custom and applicable FHA standards. Do not include items which, by established custom, are supplied by occupant and removed when he vacates premises or chattels prohibited by law from becoming realty.) _____

PORCHES:

TERRACES:

GARAGES:

WALKS AND DRIVEWAYS:

Driveway: width _____ ; base material _____ ; thickness _____ "; surfacing material _____ ; thickness _____ "

Front walk: width _____ ; material _____ ; thickness _____ ". Service walk: width _____ ; material _____ ; thickness _____ "

Steps: material _____ ; treads _____ "; risers _____ ". Cheek walls _____

OTHER ONSITE IMPROVEMENTS:

(Specify all exterior onsite improvements not described elsewhere, including items such as unusual grading, drainage structures, retaining walls, fence, railings, and accessory structures.)

LANDSCAPING, PLANTING, AND FINISH GRADING:

Topsoil _____ " thick: ☐ front yard; ☐ side yards; ☐ rear yard to _____ feet behind main building.

Lawns (seeded, sodded, or sprigged): ☐ front yard _____ ; ☐ side yards _____ ; ☐ rear yard _____

Planting: ☐ as specified and shown on drawings; ☐ as follows:

_____ Shade trees, deciduous. _____ " caliper.	_____ Evergreen trees. _____ ' to _____ ', B & B.
_____ Low flowering trees, deciduous, _____ ' to _____ '	_____ Evergreen shrubs. _____ ' to _____ ', B & B.
_____ High-growing shrubs, deciduous, _____ ' to _____ '	_____ Vines, 2-year _____
_____ Medium-growing shrubs, deciduous, _____ ' to _____ '	
_____ Low-growing shrubs, deciduous, _____ ' to _____ '	

IDENTIFICATION.—This exhibit shall be identified by the signature of the builder, or sponsor, and/or the proposed mortgagor if the latter is known at the time of application.

Date _____ Signature _____

Signature _____

FHA Form 2005
VA Form 26–1852

4

Fig. 2-1 FHA Form 2005 (Cont'd.)

SPECIFICATIONS—DIVISIONS 1-16

DIVISION 1 – GENERAL REQUIRMENTS

01010   SUMMARY OF WORK
01020   ALLOWANCES
01030   SPECIAL PROJECT PROCEDURES
01040   COORDINATION
01050   FIELD ENGINEERING
01060   REGULATORY REQUIRMENTS
01070   ABBREVIATIONS AND SYMBOLS
01080   IDENTIFICATION SYSTEMS
01100   ALTERNATES/ALTERNATIVES
01150   MEASUREMENT AND PAYMENT
01200   PROJECT MEETINGS
01300   SUBMITTALS
01400   QUALITY CONTROL
01500   CONSTRUCTION FACILITIES AND TEMPORARY CONTROLS
01600   MATERIAL AND EQUIPMENT
01650   STARTING OF SYSTEMS
01660   TESTING, ADJUSTING, AND BALANCING OF SYSTEMS
01700   CONTRACT CLOSEOUT
01800   MAINTENANCE MATERIALS

DIVISION 2 – SITEWORK

02010   SUBSURFACE INVESTIGATION
02050   DEMOLITION
02100   SITE PREPARATION
02150   UNDERPINNING
02200   EARTHWORK
02300   TUNNELLING
02350   PILES, CAISSONS AND COFFERDAMS
02400   DRAINAGE
02440   SITE IMPROVEMENTS
02480   LANDSCAPING
02500   PAVING AND SURFACING
02580   BRIDGES
02590   PONDS AND RESERVOIRS
02600   PIPED UTILITY MATERIALS AND METHODS
02700   PIPED UTILITIES
02800   POWER AND COMMUNICATION UTILITIES
02850   RAILROAD WORK
02880   MARINE WORK

DIVISION 3 – CONCRETE

03010   CONCRETE MATERIALS
03050   CONCRETING PROCEDURES
03100   CONCRETE FORMWORK
03150   FORMS
03180   FORM TIES AND ACCESSORIES
03200   CONCRETE REINFORCEMENT
03250   CONCRETE ACCESSORIES
03300   CAST-IN-PLACE CONCRETE
03350   SPECIAL CONCRETE FINISHES
03360   SPECIALLY PLACED CONCRETE
03370   CONCRETE CURING
03400   PRECAST CONCRETE
03500   CEMENTITIOUS DECKS
03600   GROUT
03700   CONCRETE RESTORATION AND CLEANING

DIVISION 4 – MASONRY

04050   MASONRY PROCEDURES
04100   MORTAR
04150   MASONRY ACCESSORIES
04200   UNIT MASONRY
04400   STONE
04500   MASONRY RESTORATION AND CLEANING
04550   REFRACTORIES
04600   CORROSION RESISTANT MASONRY

DIVISION 5 – METALS

05010   METAL MATERIALS AND METHODS
05050   METAL FASTENING
05100   STRUCTURAL METAL FRAMING
05200   METAL JOISTS
05300   METAL DECKING
05400   COLD-FORMED METAL FRAMING
05500   METAL FABRICATIONS
05700   ORNAMENTAL METAL
05800   EXPANSION CONTROL
05900   METAL FINISHES

DIVISION 6 – WOOD AND PLASTICS

06050   FASTENERS AND SUPPORTS
06100   ROUGH CARPENTRY
06130   HEAVY TIMBER CONSTRUCTION
06150   WOOD-METAL SYSTEMS
06170   PREFABRICATED STRUCTURAL WOOD
06200   FINISH CARPENTRY
06300   WOOD TREATMENT
06400   ARCHITECTURAL WOODWORK
06500   PREFABRICATED STRUCTURAL PLASTICS
06600   PLASTIC FABRICATIONS

DIVISION 7 – THERMAL AND MOISTURE PROTECTION

07100   WATERPROOFING
07150   DAMPPROOFING
07200   INSULATION
07250   FIREPROOFING
07300   SHINGLES AND ROOFING TILES
07400   PREFORMED ROOFING AND SIDING
07500   MEMBRANE ROOFING
07570   TRAFFIC TOPPING
07600   FLASHING AND SHEET METAL
07800   ROOF ACCESSORIES
07900   SEALANTS

DIVISION 8 – DOORS AND WINDOWS

08100   METAL DOORS AND FRAMES
08200   WOOD AND PLASTIC DOORS
08250   DOOR OPENING ASSEMBLIES
08300   SPECIAL DOORS
08400   ENTRANCES AND STOREFRONTS
08500   METAL WINDOWS
08600   WOOD AND PLASTIC WINDOWS
08650   SPECIAL WINDOWS
08700   HARDWARE
08800   GLAZING
08900   GLAZED CURTAIN WALLS

DIVISION 9 – FINISHES

09100   METAL SUPPORT SYSTEMS
09200   LATH AND PLASTER
09230   AGGREGATE COATINGS
09250   GYPSUM WALLBOARD
09300   TILE
09400   TERRAZZO
09500   ACOUSTICAL TREATMENT
09550   WOOD FLOORING
09600   STONE AND BRICK FLOORING
09650   RESILIENT FLOORING
09680   CARPETING
09700   SPECIAL FLOORING
09760   FLOOR TREATMENT
09800   SPECIAL COATINGS
09900   PAINTING
09950   WALL COVERING

Fig. 2-2 CSI Format for Specifications

DIVISION 10 - SPECIALITIES

0100	CHALKBOARDS AND TACKBOARDS
10150	COMPARTMENTS AND CUBICLES
10200	LOUVERS AND VENTS
10240	GRILLES AND SCREENS
10250	SERVICE WALL SYSTEMS
10260	WALL AND CORNER GUARDS
10270	ACCESS FLOORING
10280	SPECIALTY MODULES
10290	PEST CONTROL
10300	FIREPLACES AND STOVES
10340	PREFABRICATED STEEPLES, SPIRES, AND CUPOLAS
10350	FLAGPOLES
10400	IDENTIFYING DEVICES
10450	PEDESTRIAN CONTROL DEVICES
10500	LOCKERS
10520	FIRE EXTINGUISHERS, CABINETS, AND ACCESSORIES
10530	PROTECTIVE COVERS
10550	POSTAL SPECIALTIES
10600	PARTITIONS
10650	SCALES
10670	STORAGE SHELVING
10700	EXTERIOR SUN CONTROL DEVICES
10750	TELEPHONE ENCLOSURES
10800	TOILET AND BATH ACCESSORIES
10900	WARDROBE SPECIALTIES

DIVISION 11 - EQUIPMENT

11010	MAINTENANCE EQUIPMENT
11020	SECURITY AND VAULT EQUIPMENT
11030	CHECKROOM EQUIPMENT
11040	ECCLESIASTICAL EQUIPMENT
11050	LIBRARY EQUIPMENT
11060	THEATER AND STAGE EQUIPMENT
11070	MUSICAL EQUIPMENT
11080	REGISTRATION EQUIPMENT
11100	MERCANTILE EQUIPMENT
11110	COMMERCIAL LAUNDRY AND DRY CLEANING EQUIPMENT
11120	VENDING EQUIPMENT
11130	AUDIO-VISUAL EQUIPMENT
11140	SERVICE STATION EQUIPMENT
11150	PARKING EQUIPMENT
11160	LOADING DOCK EQUIPMENT
11170	WASTE HANDLING EQUIPMENT
11190	DETENTION EQUIPMENT
11200	WATER SUPPLY AND TREATMENT EQUIPMENT
11300	FLUID WASTE DISPOSAL AND TREATMENT EQUIPMENT
11400	FOOD SERVICE EQUIPMENT
11450	RESIDENTIAL EQUIPMENT
11460	UNIT KITCHENS
11470	DARKROOM EQUIPMENT
11480	ATHLETIC, RECREATIONAL, AND THERAPEUTIC EQUIPMENT
11500	INDUSTRIAL AND PROCESS EQUIPMENT
11600	LABORATORY EQUIPMENT
11650	PLANETARIUM AND OBSERVATORY EQUIPMENT
11700	MEDICAL EQUIPMENT
11780	MORTUARY EQUIPMENT
11800	TELECOMMUNICATION EQUIPMENT
11850	NAVIGATION EQUIPMENT

DIVISION 12 - FURNISHINGS

12100	ARTWORK
12300	MANUFACTURED CABINETS AND CASEWORK
12500	WINDOW TREATMENT
12550	FABRICS
12600	FURNITURE AND ACCESSORIES
12670	RUGS AND MATS
12700	MULTIPLE SEATING
12800	INTERIOR PLANTS AND PLANTINGS

DIVISION 13 - SPECIAL CONSTRUCTION

13010	AIR SUPPORTED STRUCTURES
13020	INTEGRATED ASSEMBLIES
13030	AUDIOMETRIC ROOMS
13040	CLEAN ROOMS
13050	HYPERBARIC ROOMS
13060	INSULATED ROOMS
13070	INTEGRATED CEILINGS
13080	SOUND, VIBRATION, AND SEISMIC CONTROL
13090	RADIATION PROTECTION
13100	NUCLEAR REACTORS
13110	OBSERVATORIES
13120	PRE-ENGINEERED STRUCTURES
13130	SPECIAL PURPOSE ROOMS AND BUILDINGS
13140	VAULTS
13150	POOLS
13160	ICE RINKS
13170	KENNELS AND ANIMAL SHELTERS
13200	SEISMOGRAPHIC INSTRUMENTATION
13210	STRESS RECORDING INSTRUMENTATION
13220	SOLAR AND WIND INSTRUMENTATION
13410	LIQUID AND GAS STORAGE TANKS
13510	RESTORATION OF UNDERGROUND PIPELINES
13520	FILTER UNDERDRAINS AND MEDIA
13530	DIGESTION TANK COVERS AND APPURTENANCES
13540	OXYGENATION SYSTEMS
13550	THERMAL SLUDGE CONDITIONING SYSTEMS
13560	SITE CONSTRUCTED INCINERATORS
13600	UTILITY CONTROL SYSTEMS
13700	INDUSTRIAL AND PROCESS CONTROL SYSTEMS
13800	OIL AND GAS REFINING INSTALLATIONS AND CONTROL SYSTEMS
13900	TRANSPORTATION INSTRUMENTATION
13940	BUILDING AUTOMATION SYSTEMS
13970	FIRE SUPPRESSION AND SUPERVISORY SYSTEMS
13980	SOLAR ENERGY SYSTEMS
13990	WIND ENERGY SYSTEMS

DIVISION 14 - CONVEYING SYSTEMS

14100	DUMBWAITERS
14200	ELEVATORS
14300	HOISTS AND CRANES
14400	LIFTS
14500	MATERIAL HANDLING SYSTEMS
14600	TURNTABLES
14700	MOVING STAIRS AND WALKS
14800	POWERED SCAFFOLDING
14900	TRANSPORTATION SYSTEMS

DIVISION 15 - MECHANICAL

15050	BASIC MATERIALS AND METHODS
15200	NOISE, VIBRATION, AND SEISMIC CONTROL
15250	INSULATION
15300	SPECIAL PIPING SYSTEMS
15400	PLUMBING SYSTEMS
15450	PLUMBING FIXTURES AND TRIM
15500	FIRE PROTECTION
15600	POWER OR HEAT GENERATION
15650	REFRIGERATION
15700	LIQUID HEAT TRANSFER
15800	AIR DISTRIBUTION
15900	CONTROLS AND INSTRUMENTATION

DIVISION 16 - ELECTRICAL

16050	BASIC MATERIALS AND METHODS
16200	POWER GENERATION
16300	POWER TRANSMISSION
16400	SERVICE AND DISTRIBUTION
16500	LIGHTING
16600	SPECIAL SYSTEMS
16700	COMMUNICATIONS
16850	HEATING AND COOLING
16900	CONTROLS AND INSTRUMENTATION

Fig. 2-2 CSI Format for Specifications

Guide Specifications

This Guide Specification is intended to be used as a basis for the development of an office master specification or in the preparation of specifications for a particular project. In either case, this Guide Specification must be edited to fit the conditions of use. Particular attention should be given to the deletion of inapplicable provisions. Include necessary items related to a particular project. Include appropriate requirements where blank spaces have been provided.

SECTION 06100

ROUGH CARPENTRY

PART 1—GENERAL

1.01 DESCRIPTION

A. Related Work Specified Elsewhere:
1. Forms: Section 03150.
2. Finish Carpentry: Section 06200.
3. Heavy Timber Construction: Section 06130.
4. Prefabricated Structural Wood: Section 06170.
5. Architectural Woodwork: Section 06400.
6. Gypsum Wallboard: Section 09250.
7. Painting: Section 09900.

B. Work Installed but Furnished by Others:
1. Steel splice plates in Structural Metal Framing: Section 05100.
2. Bearing plates in Structural Metal Framing: Section 05100.

1.02 QUALITY ASSURANCE

A. Lumber Grading Rules and Wood Species to be in conformance with PS 20.

B. Grading rules of following associations apply to materials furnished under this section:

> 1.02.B. Check for availability of species in project locality before specifying.

1. Northeastern Lumber Manufacturer's Association, Inc. (NELMA).

> 1.02.B.1 Includes Red or White Spruce (Eastern Spruce) and Eastern White Pine (Northern White Pine).

2. Southern Pine Inspection Bureau (SPIB).
3. West Coast Lumber Inspection Bureau (WCLIB).

> 1.02.B.3 Includes Douglas Fir, Western Cedar, Mountain Hem-Fir, and Sitka Spruce.

4. Western Wood Products Association (WWPA).

> 1.02.B.4 Includes Douglas Fir, Douglas Fir (South), Hem-Fir, Western Larch, Mountain Hemlock, Idaho White Pine, Lodgepole Pine, Ponderosa Pine, Sugar Pine, Western Red Cedar, Englemann Spruce, Subalpine Fir, and Incense Cedar, Western Hemlock.

5. Redwood Inspection Service (RIS).

October 1978 06100/3

9. Western Wood Products Association (WWPA):
  a. Standard Grading Rules for Western Lumber, 1977.

1.03 SUBMITTALS

A. Samples:
1. Submit samples of wood decking which will be exposed in finished work, to show face texture and color of material.

> 1.03.A.1 Delete if wood decking is not required.

B. Shop Drawings:
1. Submit shop drawings indicating framing connection details, fastener connections, and dimensions.

> 1.03.B.1 Delete when details are included on architectural or structural drawings.

2. Indicate pattern of wood decking members.

> 1.03.B.2 Delete if wood decking is not required.

C. Certification:
1. Pressure treated wood: Submit certification by treating plant stating chemicals and process used, net amount of salts retained, and conformance with applicable standards.
2. Preservation treated wood: Submit certification for water-borne preservative that moisture content was reduced to 19% maximum, after treatment.
3. Fire-retardant treatment: Submit certification by treating plant that fire-retardant treatment materials comply with governing ordinances and that treatment will not bleed through finished surfaces.

1.04 PRODUCT DELIVERY, STORAGE, AND HANDLING

A. Immediately upon delivery to job site, place materials in area protected from weather.

B. Store materials a minimum of 6 in. (150 mm) above ground on framework or blocking and cover with protective waterproof covering providing for adequate air circulation or ventilation.

C. Do not store seasoned materials in wet or damp portions of building.

D. Protect fire-retardant materials against high humidity and moisture during storage and erection.

E. Protect sheet materials from corners breaking and damaging surfaces, while unloading.

PART 2—PRODUCTS

2.01 MATERIALS

A. Lumber:
1. Dimensions:
  a. Specified lumber dimensions are nominal.
  b. Actual dimensions to conform to PS 20.
2. Surfacing: Surface four sides (S4S), unless specified otherwise.
3. End jointed lumber:
  a. Structural purposes interchangeable with solid sawn lumber.
  b. Glued joints of loadbearing lumber: PS 56.

6/06100 October 1978

PART 3—EXECUTION

3.01 INSPECTION

A. Verify that surfaces to receive rough carpentry materials are prepared to require grades and dimensions.

3.02 INSTALLATION

A. Sills:

> 3.02.A Use treated lumber any commercial softwood species or all heart wood western cedar or redwood.

1. Set level 1/16 in., (1.6 mm) in 6 ft., (1.8 m) in mortar bed, 1 in. (25 mm) from exterior face of foundation.
2. Secure sills with 1/2 in. x 8 in. (13 mm x 203 mm) minimum size anchor bolts embedded in the structure minimum of 6 in. (152 mm), spaced maximum of 4 ft. (1.2 m) o.c.
3. Join solid sill members with halved joints, where not continuous and at corners, minimum of 1 ft. (0.3 m) lapped joint.
4. Lap built-up sill members minimum distance of 2 ft. (0.6 m).
5. Termite shields:
  a. Bed termite shield in mortar bed and extend across top of foundation wall, bend down 2 in. (51 mm) at angles of 45 degrees both inside and out.

> 3.02.A.5.a Use full termite shields over masonry foundation wall.

** OR **

  a. Bed termite shield in mortar bed and extend half distance across top of foundation wall, bend down 2 in. (51 mm) at angle of 45 degrees on outside.

> 3.02.A.5.a Use half termite shield over concrete foundation wall.

* * * * *

  b. Fit around anchor bolts and fill joints with asphaltic mastic or solder.

B. Posts or Columns:
1. Provide two surfaces on posts at right angles to each other for installation of interior finish materials.
2. Built-up posts: Arrange and nailed together to accommodate type of construction.
3. Provide mortise in posts to receive tenon connections of girts.

> 3.02.B.3 Mortise and tenon connections for post construction only. Delete if not required.

4. Erect posts straight, plumb with straight edge and level, and brace with tack boards at plate and sill.

C. Girts:
1. Install continuous girts, running from post to post.

October 1978 06100/13

  d. Nail 6 in. (152 mm) o.c. along panel edges and 12 in. (0.3 m) o.c. at intermediate supports.
  e. Nail 6 in. (152 mm) o.c. at all supports, for support spaced 4 ft. (1.2 m) o.c.
  f. Use 6d common, smooth ring-shank, or spiral-thread nails for panels 1/2 in. (13 mm) thick or less and 8d for greater thickness, except that when panels are 1 1/8 in. (28 mm) or 1 1/4 in. (31 mm) use 8d ring-shank or 10d common.

K. Wall Sheathing:
1. Board sheathing:
  a. Install _____ with end joints staggered and terminated on supports.

> 3.02.K.1.a (diagonally); (horizontally).

  b. Nail at each bearing two 8d nails driven to full penetration.
  c. Cover wood sheathing with sheathing paper applied horizontally in "shingle" fashion starting at bottom, minimum lap of 4 in. (102 mm).
  d. Secure sheathing paper using roofing nails with metal nailing discs spaced 12 in. (0.3 m) on center vertically and horizontally, starting at each horizontal lap.

2. Plywood sheathing:
  a. Install with face grain horizontal or vertical.
  b. Allow minimum 1/16 in. (1.6 mm) space at end joints and 1/8 in. (3.2 mm) at edge joints, doubling these spacings in wet or humid conditions.
  c. Nail 6 in. (152 mm) o.c. along panel edges and 12 in. (305 mm) o.c) at intermediate supports with 6d common nails for panels 1/2 in. (13 mm) thickness and 8d nails for greater thickness.

L. Subflooring:
1. Board subflooring:
  a. Lay diagonally at 45° angle to joists.
  b. Cut ends parallel to joists and terminate over supports.
  c. Depress for ceramic tile and nail own cleats to side of joists.

> 3.02.L.1.c Delete when not applicable.

  d. Support edges with wood blocking.

** OR **

  d. Tongue and groove material:
    (1) Nail each board at each bearing point using one 8d nail for each 2 in. nominal face width of board.
    (2) Drive one nail through tongue and other nail(s) through face.
    (3) Drive nails to full penetration.

* * * * *

2. Plywood subflooring:
  a. Install with face grain perpendicular to joists; end joints occurring over the joists.
  b. Allow 1/16 in. (1.2 mm) space at end joints and 1/8 in. (3.2 mm) at edge joists.
  c. Stagger panel end joists.
  d. Nail subflooring 6 in. (152 mm) on center along panel edges and 10 in. (254 mm) on center at intermediate supports with 6d common nails for 1/2 in. (12 mm) plywood and 8d nails for greater thickness.
  e. Nail subflooring with 8d ring-shank or 10d common nails, spaced 6 in. (152 mm) on center at all supports for 1 1/8 in. (28 mm) or 1 1/4 in. (31 mm) thick panels, and supports are 4 ft. (1.2 m) on center.

October 1978 06100/19

Fig. 2-3 Sample pages from the CSI Specification Series: Rough Carpentry

It is extremely important that specifications be studied with great care. The misunderstanding of a single word or phrase can lead to legal difficulties, as well as loss of time or money to the contractor. It is also important that plans and specifications be read together so that a full picture of the building which the architect envisions is obtained.

Specification Formats

There are several different formats that can be followed in writing specifications. Two of the most commonly used are from the Federal Housing Administration (FHA) and the Construction Specifications Institute (CSI). The FHA form, figure 2-1, is a four-page document that is divided into twenty-seven sections. Each section represents a particular construction phase, listed in sequence as it occurs on the job.

The CSI format lists sixteen phases of construction, figure 2-2. Figure 2-3 illustrates several sample pages from the section entitled "Rough Carpentry." The complete form is much longer.

When filling out specification forms, it is important to use simple, direct language, give a good description of the materials, and use accepted standards.

ASSIGNMENT

A. Answer the following questions, referring when necessary to the specifications in the Appendix.

1. Who is responsible for clearing the site for construction?

2. What is the purpose of specifications?

3. If a conflict regarding a specific item arises between specifications and working drawings, which of the two is followed?

4. Vertical siding is made of what material?

5. Who furnishes the mantel in the living room?

6. What kind of wood is used to make the treads of the cellar stairs?

7. What precaution must be taken before applying dampproofing to surfaces?

8. How thick is the gravel under concrete floors?

9. What are the proportions required for cement mortar?

10. What standard specifications are used to specify the type of concrete block wanted?

11. The fire brick to be used must meet what standard specifications?

12. The fireplace in the recreation room is faced with what kind of brick?

13. What is the thickness of the setting bed under the flagstone?

B. Answer the following questions, giving the article and paragraph of the specifications in which the information was located.

1. How is trellis work fastened to concrete or stone work?

2. What material is used to cover the walls of the lavatory and workshop?

3. What kind of face brick is called for in the fireplace in the living room?

4. What covering is called for on the countertop of the kitchen cabinets?

5. What type of material is used for the finish floors?

UNIT 3 SCALE AND WORKING DRAWINGS

OBJECTIVES

- Describe a scale drawing.
- List and define the various drawings in a set of plans.
- Read an architect's scale.

SCALE DRAWINGS

Working drawings, or blueprints, are used by most construction workers to show how the various parts of the project fit together. The drawings are usually shown on 24-inch by 36-inch paper. This means, of course, that the features are much smaller on paper than their actual size. However, the features are always drawn in proportion to the other features on the working papers and in proportion to the size of the actual construction project. This is called *drawing to scale.*

To illustrate, if a scale of 1 inch equals 1 foot (1″ = 1′-0″) is used, a line 6 inches long on the drawing represents a line that is actually 6 feet long.

The scale which most drafters use is laid out so that each edge contains two different scales. Figure 3-1 shows that the 1/8-inch and

1/4-inch scales are on one edge, while the 1/2-inch and the 1-inch scales are on the other. The 1/8-inch scale is marked off with short lines. Between each short line is a longer line. The distance between each short and long line represents 1 actual foot. The scale is marked off with longer lines. The distance between each long line on this scale represents 1 foot. Each scale has a short section at the end with finely calibrated lines. These fine graduations represent inches or fractions of inches in the scale being used, figure 3-2.

While a drafter uses the architect's scale to lay out drawings, it should never be used by a carpenter to scale a drawing for a dimension. This is for several reasons. The paper could have changed in shape or size, or a mistake could have been made by the carpenter or drafter. If the drawing is not properly

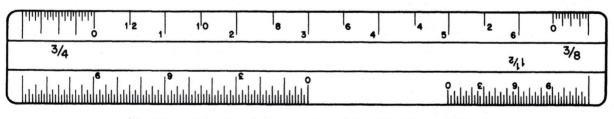

Fig. 3-1 Architect's scale representing 1/8″, 1/4″, 1/2″, and 1″ scales

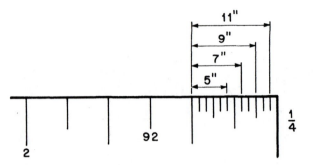

Fig. 3-2 Fine calibrations represent inches or fraction
of inches on the scale.

dimensioned, the drafter, contractor, or architect is consulted.

WORKING DRAWINGS

A set of *working drawings* contains information in the form of drawings, notes, dimensions, and other indications necessary for the construction of a building. There are several different types of drawings that are contained in a set of working drawings. The more common ones are plot plans, foundation plans, floor plans, sections, detail drawings, and elevations.

Plot Plan

A *plot plan* is used to show how the building is located in relationship to the property line, figure 3-3. The plan is drawn from a position looking directly at the top of the building. No details are shown on a plot plan. They represent large areas and are drawn to a relatively small scale, usually about 1" = 20'-0".

Foundation Plan

A *foundation plan* is drawn from a position looking straight down on the building's foundation, figure 3-4. Foundation plans show the general shape of the foundation and the location of foundation walls, pilasters, columns, piers, and footings. Most foundation plans are drawn to the scale of 1/4" = 1'-0".

Floor Plan

A *floor plan* also is drawn from a position looking straight down on the building, figure 3-5. To draw it, a theoretical cut is made through the entire building on a horizontal plane (plane parallel to the ground). The building above the cut line is then removed. This allows an unobstructed view looking directly down at the building.

Most floor plans are drawn to a relatively small scale, usually about 1/4" = 1'-0". The scale used usually depends upon the size of the drafting paper and the size of the foundation.

Sections

A *sectional view* is drawn by making a theoretical vertical cut through a portion of the structure, figure 3-6. It is done in an area that needs to be shown in greater detail. Sectional views are mainly drawn to show the relationship of various construction members to each other. They are drawn to a relatively large scale, usually 3/4" = 1'-0".

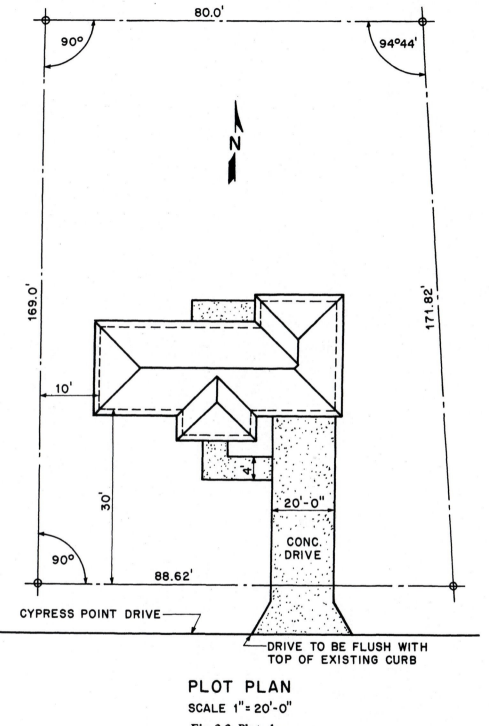

80.0'

90°

94°44'

N

169.0'

171.82'

10'

30'

4'

20'-0"

CONC.
DRIVE

90°

88.62'

CYPRESS POINT DRIVE

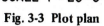

DRIVE TO BE FLUSH WITH
TOP OF EXISTING CURB

PLOT PLAN
SCALE 1"= 20'-0"

Fig. 3-3 Plot plan

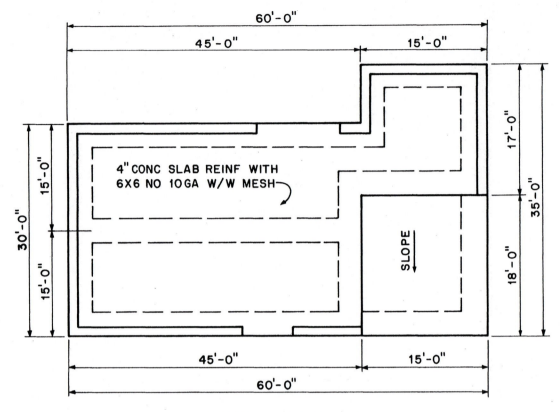

FOUNDATION PLAN

SCALE $\frac{1}{4}" = 1'-0"$

Fig. 3-4 Foundation plan

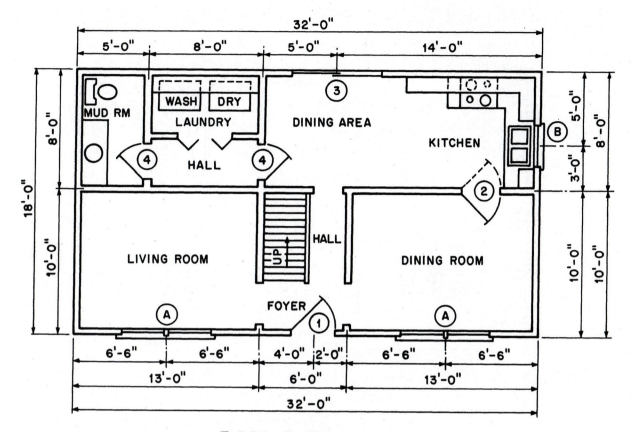

FIRST FLOOR PLAN

SCALE $\frac{1"}{4}$=1'-0"

Fig. 3-5 Floor plan

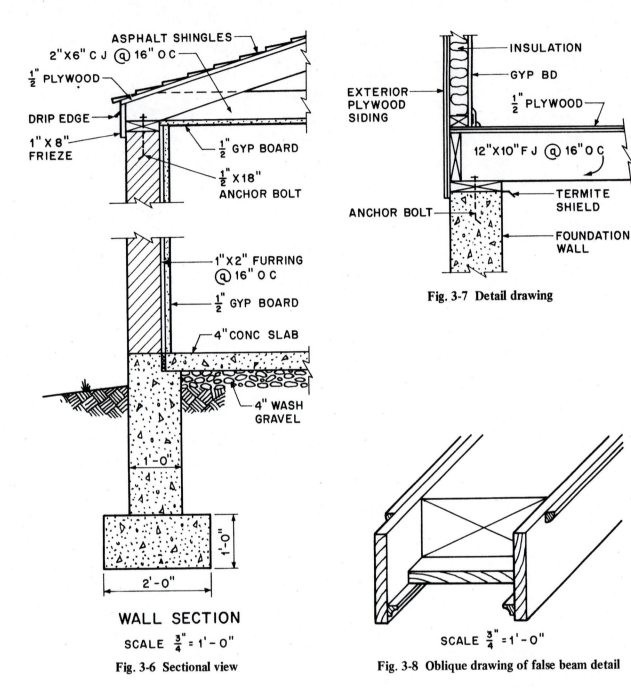

ASPHALT SHINGLES

2"X6" C J @ 16" O C

$\frac{1}{2}$" PLYWOOD

DRIP EDGE

1" X 8" FRIEZE

$\frac{1}{2}$" GYP BOARD

$\frac{1}{2}$" X 18" ANCHOR BOLT

1"X 2" FURRING @ 16" O C

$\frac{1}{2}$" GYP BOARD

4" CONC SLAB

4" WASH GRAVEL

1'-0"

1'-0"

2'-0"

WALL SECTION

SCALE $\frac{3}{4}$" = 1'-0"

Fig. 3-6 Sectional view

INSULATION

GYP BD

$\frac{1}{2}$" PLYWOOD

EXTERIOR PLYWOOD SIDING

12"X10" F J @ 16" O C

TERMITE SHIELD

ANCHOR BOLT

FOUNDATION WALL

Fig. 3-7 Detail drawing

SCALE $\frac{3}{4}$" = 1'-0"

Fig. 3-8 Oblique drawing of false beam detail

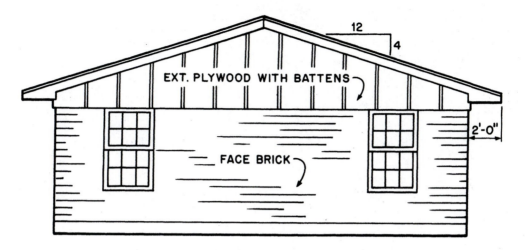

LEFT-SIDE ELEVATION

SCALE $\frac{1}{4}'' = 1'-0''$

Fig. 3-9 Left-side elevation

Detail Drawings

Detail drawings show the construction assembly of various members, parts, or items, figure 3-7. Most details are drawn using only two dimensions, height and width. But sometimes details are drawn in isometric or oblique form. *Isometric drawings* are a three-dimensional view. The object is turned and tilted so that all three faces appear equal. *Oblique drawings,* figure 3-8, are also three-dimensional views. One face is parallel to the picture plane and receding lines are parallel to each other.

Elevations

An *elevation* is a two-dimensional drawing, figure 3-9. It is drawn as though the observer were looking straight at a side of the building. There are two methods of designating elevation drawings. In the first method, they are labeled front, rear, left side, and right side. In the second method, the elevations are labeled according to the compass direction in which they face. If the front of the house faces south, this is called the south elevation; the right side the east elevation; the left side the west elevation; and the rear, the north elevation. Elevations are drawn to a relatively small scale, usually 1/4" = 1'-0".

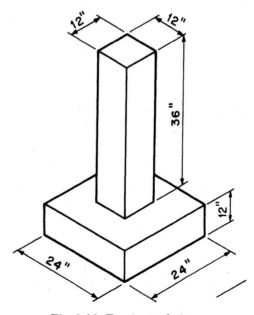

Fig. 3-10 Footing and pier

ASSIGNMENT

A.   Give the missing dimensions in the following illustrations of the architect's scale.

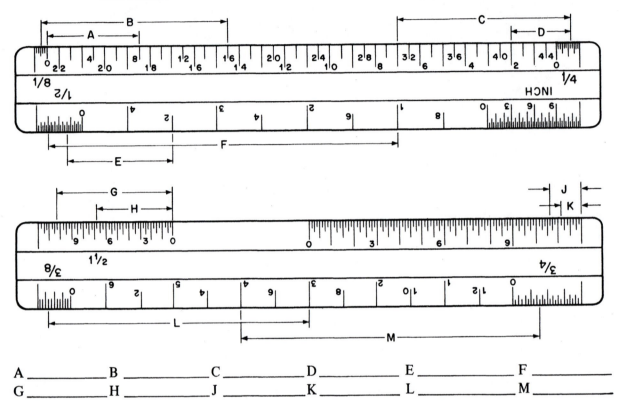

A _____   B _____   C _____   D _____   E _____   F _____

G _____   H _____   J _____   K _____   L _____   M _____

B.   Provide the correct word or phrase to complete the statement or answer the question.

1. A plan view is one which is seen as if a _____ cut has been made through the object.

2. A plan view differs from a sectional view because in the sectional view the cut has been made _____.

3. When a drawing is made smaller than the size of the actual object but is drawn in proportion to the object, this is called drawing to _____.

4. The 1/8-inch scale is read from _____ to _____.

5. If a drawing is not properly dimensioned, the _____ or _____ is consulted.

6. Site plans are usually drawn to a scale of _____.

7. Most details are drawn using two dimensions: _____ and _____.

8. There are usually _____ elevations in every set of working drawings.

9. Sectional drawings used in construction of buildings are usually drawn to a scale of _____.

10. Each edge of the architect's scale contains _____ different scales.

11. A plot plan shows the location of the building in relationship to the
    _____.

12. Most foundation plans are drawn to a scale of _____.

13. An elevation is a (an) _____-dimensional drawing.

14. What is the name of the street in the plot plan in figure 3-3? _____

15. In what direction does the house face in figure 3-3? _____

C. Sketch an elevation of the pier
    and footing in figure 3-10.

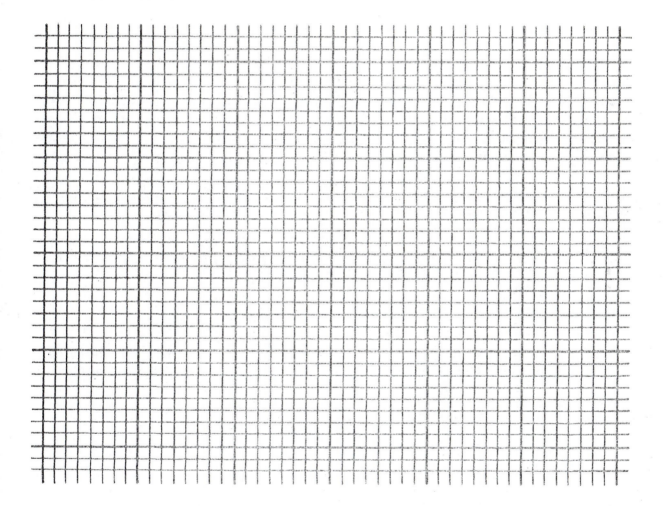

UNIT 4 THE PLOT PLAN

OBJECTIVES

- Describe the importance of plot plans.
- State what type of information is generally shown on plot plans.
- Read plot plans.

Plot plans, sometimes called *site plans*, show the location of the building, drives, and sidewalks in relationship to property lines, figure 4-1. Plot plans also show details such as existing trees that are to be removed and those that are to remain and utility service. The location, elevation, and size of the plot in relation to certain surface data are also indicated. Because of this, plot plans are important to the building designer, estimator, contractor, and owner.

In plot plan terminology, the terms *location, elevation,* and *size* have distinct meanings.

- *Location* of land is determined by a surveyor from city or town land records and from stone monuments or other markers established by other surveyors, private or government.

- *Elevation* of land is determined from a known coastal sea level. Almost every town or city has one or more permanent points established from this known sea level. These points, from which local land elevations are taken, are known as county or city *datum.* The points of established

elevations are sometimes found on the walls of permanent buildings or are marked by metal rods in fixed concrete bases set in the ground. These are called *bench marks.*

- *Size* of the plot is determined by its boundary lines. The lines on a survey indicate the length and compass direction of each boundary line.

In some areas, it may not be necessary to have a plot plan because of the absence of property restrictions. In most areas, however, zoning laws place restrictions and requirements on the type, size, and location of buildings and on building materials. These restrictions vary from one locality to the next. Most lending agencies require a survey or plot plan be submitted before loans are approved. The purpose of these restrictions is to protect the residential property owner.

Plot plans are drawn up by registered surveyors or engineers.

Contour Lines

When the surface or profile of the plot of ground is irregular, the survey shows contour lines on the plot plan, figure 4-2. *Contour lines* are continuous lines extending to points of the same elevation across the surface of a plot of ground.

The distance between contour lines is called the *vertical contour interval.* Contour intervals of 1 or 2 feet are sometimes used for fine grading, but intervals of 3 to 5 feet are usually used. If contour lines are set up with 2-foot intervals, a 2-foot slope occurs regardless of the distance between the contour lines. Therefore, the closer together the lines are,

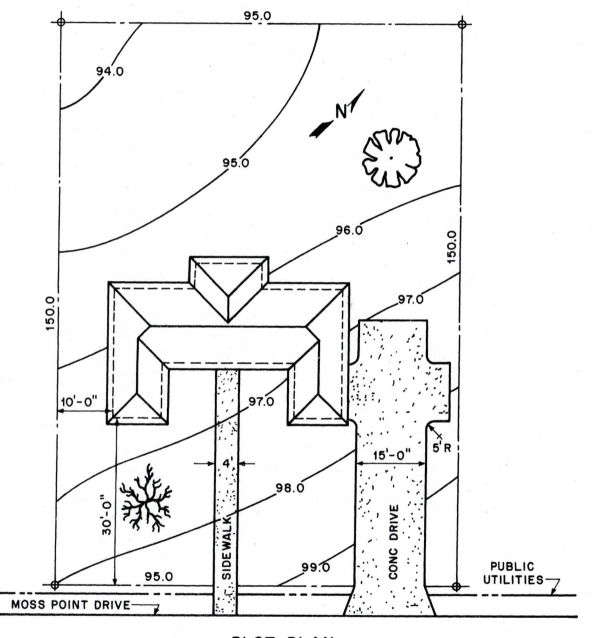

PLOT PLAN
SCALE 1"=20'-0"
Fig. 4-1 Plot plan

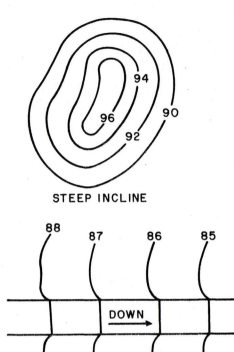

STEEP INCLINE

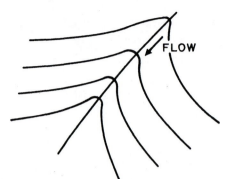

FLOWING STREAM

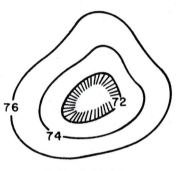

ROAD   DEPRESSION

Fig. 4-2 Contour lines

the steeper the slope is at that point. The farther apart the lines are, the more gradual the slope.

Most contour line elevations are established by referring to an arbitrary bench mark that has an assigned elevation. Any elevation can be used for the bench mark, but 100 is often used. This does not mean that the elevation of the bench mark is 100 feet above sea level. It is simply used as a reference for the contour lines.

Property Lines

Property lines are usually represented by centerlines that enclose the plot. The dimensions of the line are placed on that line. The property line can be dimensioned in feet and inches, or feet and decimal parts of a foot. To measure the property line in feet and decimal parts, an engineer's tape is used. The tape

shows each foot divided into ten parts (0.1 foot) and each tenth divided into ten more parts (0.01). The measurements, such as 112.129, represent decimal feet.

To convert such a dimension into feet and inches, the decimal feet measurement is multiplied by 12. This gives the number of inches and decimal inches. Then, the remainder is multiplied by 16, 32, or 64 to find the number of 16ths, 32nds, or 64ths.

Example: Convert 112.129 feet to feet and inches (to the closest 64th inch).

0.129 x 12 = 1.548 inches

0.548 (remainder) x 64 = 35.072 = $\frac{35}{64}$

Answer: 112 feet 1 $\frac{35}{64}$ inches

The direction of property lines is given in one of two ways. Sometimes the angle

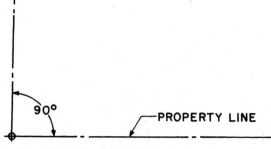

Fig. 4-3 Indicating angle between property lines

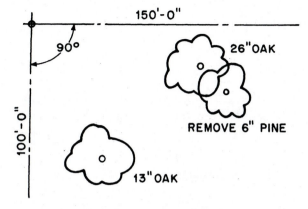

Fig. 4-4 Noting trees

between two lines is given, figure 4-3. Other times the bearing of the line is given. A *bearing* is the direction of a line measured in degrees and minutes from north, east, south, or west. A typical bearing is S65°-10'W. This means the line is 65 degrees 10 minutes west of due south.

Trees

Trees are usually drawn and noted on the plot plan, figure 4-4. Usually, the diameter and species of tree is noted, such as *18" oak*. If the tree is to be removed, the note might read *remove existing 18" oak*. Sometimes, the note simply reads *existing to be removed*.

Location of Buildings

How a building is located on the lot usually depends on zoning laws or guidelines established by lending agencies. Most building setbacks are a minimum of 20 feet from the front property line and 5 feet from the property lines on each side. These are usually the only dimensions that are required to locate

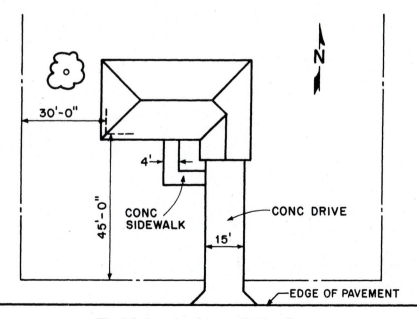

Fig. 4-5 Locating drives and sidewalks

the building. A *setback* is the distance from the property line to the building.

Drives and Sidewalks

Drives and sidewalks are drawn in and dimensioned on the plot plan, figure 4-5.

Necessary dimensions usually include the width and any necessary radii. Expansion joints are represented by a single line. A note is then used to give a description of the expansion joint, such as *1″ x 4″ redwood expansion joints @ 10′-0″ OC.* Any special finishing techniques required on the drive also are noted.

ASSIGNMENT

Answer the following questions, referring to the plot plan on Sheet 1/7 in the back of the book.

1. What are the dimensions of the lot?

2. To what scale is the plot plan drawn?

3. What is the dimension from the property line to the front of the house?

4. What is the dimension from the sidewalk to the property line?

5. What is the area of the lot?

6. If the drainage field is 2 feet below the 93 contour line, how many feet below the sidewalk is it?

7. Give the approximate number of square feet of flagstone needed for the front walk.

8. Approximately how many feet of pipe are needed from the house to the septic tank?

9. Approximately how many feet of pipe are needed from the septic tank to the field?

10. What is the source of the water supply for this house?

11. What is the approximate pitch in feet from the back of the house to the rear property line?

12. In what direction does the house face?

13. How close is the septic tank to the flagstone terrace?

14. What are the dimensions of the garden plot?

15. What is the area of the garden plot in square yards?

16. According to the location of the house on the plot plan, where is the most gradual slope in the finish grade located?

17. In accordance with the compass direction on the plot plan, identify the direction of each corner of the plot.

18. What does the least distance between the contour lines indicate?

19. Where does the steepest slope in the finish grade occur?

20. What is the approximate difference in elevation between the front and the rear of the building along the centerline?

21. According to the requirements of the finish, is it necessary to remove excess soil from the property?

22. Briefly, what is the major function of contour lines?

23. What determines the spacing of contour lines?

24. Give the equivalent in standard feet, inches, and fractions of an inch of the engineer's elevation of 136.39 feet to the nearest 16th of an inch.

25. A bench mark elevation is 132.64 feet and the bottom of the foundation footing is 121.29 feet. Find the difference between these measurements in feet, inches, and fractions of an inch to the nearest 16th of an inch.

26. Sketch the plan symbol of the cupola.

27. The plan view of the roof shows that the residence has what type of roof?

28. Does this plan view of the roof indicate any overhang? Explain your answer.

UNIT 5   THE FOUNDATION PLAN

OBJECTIVES

- Interpret the foundation plan.
- Recognize how footings, columns, and joists are indicated on plans.
- Read the usual methods of dimensioning.

Buildings are supported by the *foundation*. Without a properly built and designed foundation, the structure settles, the walls are out of square, finish materials crack, and doors and windows do not operate properly.

There are three basic types of spread foundations used in light construction. They are slab-on-grade foundations, crawl-space foundations, and basements.

SLAB-ON-GRADE FOUNDATIONS

A slab-on-grade, figure 5-1, is a popular foundation in southern states in the United States. It has been used successfully

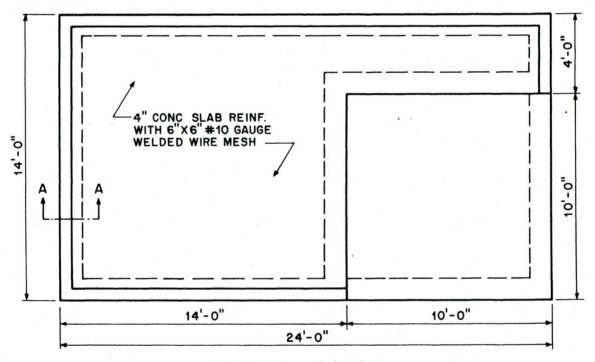

Fig. 5-1 Slab-on-grade foundation

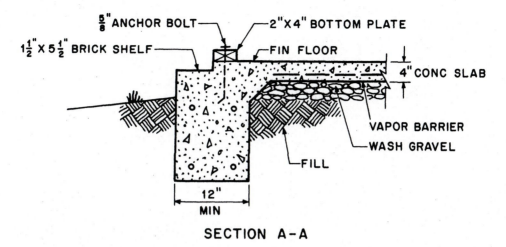

SECTION A-A

Fig. 5-2 Footing detail

nationwide. The most popular type of slab-on-grade foundation system has a 4-inch slab that is turned down on the edge to create a footing, figure 5-2.

The foundation plan shows the overall size of the building, location of footings, and any depressions that might appear in the slab.

The footings are indicated by a series of dashed lines, usually located 12 inches from the outside edge of the foundation. Dashed lines are a standard indication of footings in plan view. This line varies, however, depending upon soil conditions and code requirements. If there are any interior footings, they are indicated

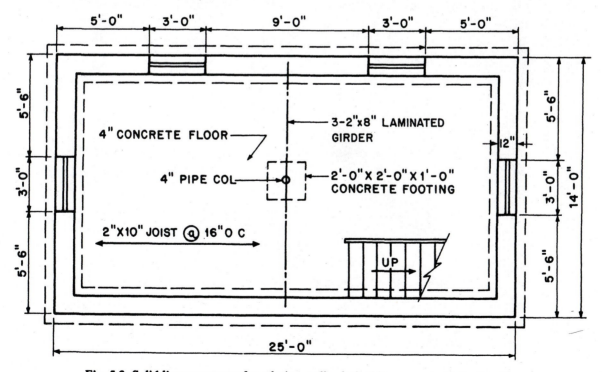

Fig. 5-3 Solid lines represent foundation walls; dashed lines represent the footing

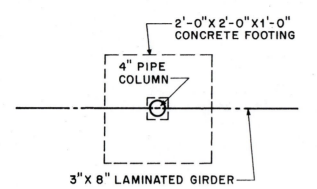

Fig. 5-4 Pipe column, footing, and girder indicators

$$2" \times 10" \text{ F J } @ 16" \text{ O C}$$

Fig. 5-5 Joist indicator

by two parallel dashed lines approximately 12 inches apart. If the structure is brick veneer, there is a solid line 5 1/2 inches in from the line that represents the outside edge of the foundation.

BASEMENTS AND CRAWL-SPACE FOUNDATIONS

Basements and crawl-space foundations are usually constructed of load-transmitting elements such as piers, columns, foundation walls, and pilasters. The loads are transmitted to concrete footings that spread them over a given area. The foundation walls are represented by solid lines. Dashed lines on each side of the foundation wall represent the footing, figure 5-3. Figure 5-4 shows how to indicate a column, its footing, and the girder above the column.

Joists

Another feature usually found on the basement plan is the joist indicator. This shows the direction of the first-floor joist with a note stating the size and distance between each joist. Figure 5-5 illustrates this indicator with a note *2" x 10" FJ*. The arrow shows the direction in which the joist

will run. The note *16" OC* specifies that the joists are to be 16 inches apart, center to center.

Joists indicated on a basement or crawl-space foundation plan are the first floor joists. They are overhead when standing on the basement floor. Similarly, the floor plan of the first floor shows second floor joists; the plan of the second floor shows third floor joists; and so on.

Foundation Walls and Partitions

The methods of dimensioning foundation walls and partitions are fairly well standardized. Figure 5-6 shows the usual methods of indicating and dimensioning walls and partitions of various materials on foundation walls or in any horizontal sectional view.

Areaways

An *areaway* is a sunken space usually located in front of, or near, a basement door or window. It is designed to provide natural lighting, ventilation, and access to a door or window.

Areaways are usually set below grade. They are made of masonry, semirustproof sheet metal, or galvanized iron. Better construction sometimes has a concrete bottom pitched to a drain. Other types have a cinder or gravel bottom to absorb excess water. Figure 5-7 illustrates a concrete rectangular areaway. A metal semicircular areaway is shown in figure 5-8. Both are drawn as they would appear on a basement plan.

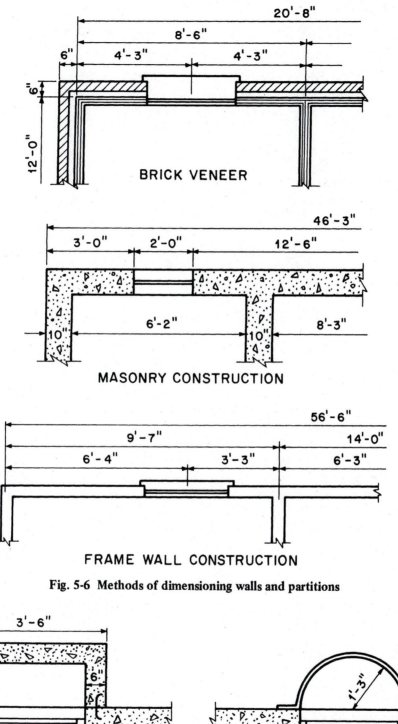

BRICK VENEER

MASONRY CONSTRUCTION

FRAME WALL CONSTRUCTION

Fig. 5-6 Methods of dimensioning walls and partitions

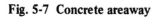

PLAN

SCALE $\frac{3}{4}$" = 1'-0"

Fig. 5-7 Concrete areaway

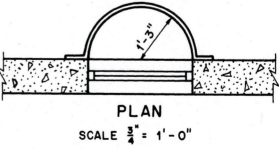

PLAN

SCALE $\frac{3}{4}$" = 1'-0"

Fig. 5-8 Metal areaway

ASSIGNMENT

Answer the following questions, referring to the basement plan of the house in figure 5-3.

1. What are the overall dimensions of the building?

2. Is the building a square, a hexagon, or a rectangle?

3. How many windows are shown?

4. Are there any doors in this basement?

5. How thick are the foundation walls?

6. Is the height of the walls shown?

7. Is there a footing under the wall?

8. Are the windows at the rear of the house the same distance from the corners?

9. What is the size of the girder?

10. What is the dimension from the center of the girder to the outside face of the right-hand wall?

11. How many girders are shown?

12. What is the area of the cellar floor?

13. If the cellar floor is 4 inches thick, how many cubic feet of concrete are needed for it?

14. What is the outside perimeter of the cellar wall?

15. What size pipe column is used?

16. How large is the footing for the pipe column?

17. Is the girder in the center of the house?

18. Do the floor joists run at right angles or parallel to the girder?

19. What is the inside perimeter of the cellar wall?

20. What distance are the joists placed from center-to-center?

UNIT 6 THE FLOOR PLAN

OBJECTIVES

- Explain the function of bearing partitions, nonbearing partitions, and opening supports.

- Describe the floor plan of the residential drawings in terms of framing construction.

- Identify indicators in plan views not shown as standardized symbols.

WOOD PARTITIONS

Wood partitions have two main structural functions. They support the floor and roof framing above them, and they form the exterior and interior walls of the rooms of the building. In general there are two classifications: bearing and nonbearing.

Bearing Partitions

Load-bearing partitions form a continuous line of support from the foundation or girder up through the building to the joist or rafter bearings. They are never less than 3 1/2 inches wide, have double upper plates, and have studs which are spaced so that they are located directly underneath the joists which they support, figure 6-1.

Openings for arches, doors, and supply ducts often interfere with the regular spacing of studs in bearing partitions. Therefore, these openings must be reinforced at the heads by increasing the depth of the wood headers, forming truss work, or including steel lintels to carry the joists above the openings, figure 6-2.

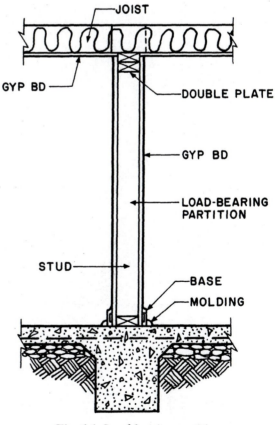

Fig. 6-1 Load-bearing partition

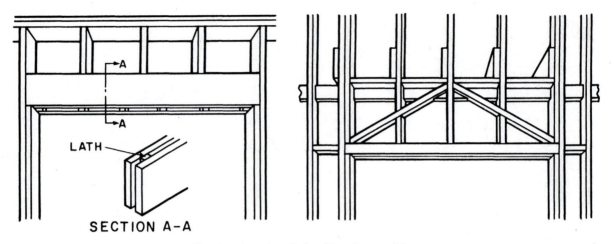

Fig. 6-2 Openings in load-bearing partitions

Nonbearing Partitions

Nonbearing partitions carry no load from the upper parts of the building. They divide the floor space into rooms and provide a surface on which trim and wall coverings are fastened. They also supply support and enclosure for heating and plumbing lines and ducts. This type of partition may run parallel to, or at an angle to, the floor joist.

Nonbearing partitions are 3 1/2 inches in width and are tied to the load-bearing partitions by means of the cap plate, tees, and corners.

TRADE SYMBOLS AND INDICATORS

Symbols

Architectural symbols are well standardized throughout the building industry. However, architects sometimes provide their own interpretation of the symbols and abbreviations in the drawing. The carpenter must always be careful of those symbols which are unfamiliar. It is very important that any notes appearing on the drawings be fully understood.

The carpenter must have a thorough knowledge of the trade symbols for cabinet, masonry, plumbing, heating, and electrical work. This helps the carpenter understand the project as a whole, as well as relate these fields specifically to the carpentry work. For example, the carpenter must understand how a discharge line runs from a particular plumbing fixture and the symbols used to express this feature.

In many cases, a complete listing of trade symbols are not shown on the set of drawings made for the construction contractor. It is then important to inspect the plumbing, heating, and electrical drawings if a detailed description and the location of such symbols are required.

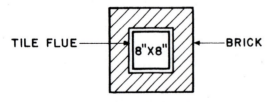

Fig. 6-3 Plan single-flue chimney

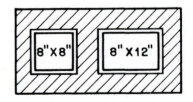

Fig. 6-4 Plan double-flue chimney

Indicators

When reading a plan view, it is important to note where the horizontal cut is made or what floor level that particular drawing represents.

For example, the construction of a chimney varies from floor level to floor level. Figures 6-3 and 6-4 illustrate a brick chimney with one and two flues respectively. Each heating element, furnace, and fireplace requires a separate flue, and these elements are frequently on different floor levels. It can readily be seen, then, that the number of flues and their arrangement in the chimney vary at different heights.

Figure 6-5 shows a one-flue chimney as seen on the basement plan. Figure 6-6 shows the same chimney on the first-floor plan.

Some indicators used to show doors and windows are illustrated in figure 6-7. These indicators are basic, although there are as many different types of indicators as there are types of windows and doors. Some of these other indicators and conventions are discussed in later units.

Stairways are of a variety of designs. They may be constructed of either wood, brick, or concrete. Figure 6-3 illustrates several different types of stair plans.

The drawings in figure 6-8 represent complete flights of stairs. Such is not the case on floor plans, however. Since a floor plan is a horizontal section, the stairs are "cut" by the horizontal plane, and only half a flight is visible at any given floor level.

Figure 6-9 shows a typical flight of basement stairs as they appear on the basement plan, Part A. On the first-floor level, these stairs would appear as they do in Part B.

Frequently in building construction, a flight of stairs is built directly over a flight of stairs on the floor below. This is considered good practice as it saves space and gives better headroom. Figure 6-10 illustrates such construction in section and in plan views.

Walls on the floor plan are indicated by a series of parallel lines. Interior partitions are usually indicated by lines that are 3 1/2 inches apart. The distance between the lines that represent the exterior wall varies with the type of material used.

Most of the dimensions on a floor plan are given on the outside of the outline of the building, figure 6-11. The dimension line nearest the outline of the building usually gives the location of partitions, corners, and windows. The second dimension line usually gives only the distance between partitions. The outside dimension line gives an overall dimension. Dimensioning practice varies from architect to architect, but they all use the same basic approach.

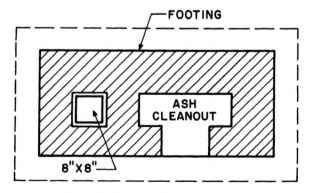

Fig. 6-5 Chimney foundation plan

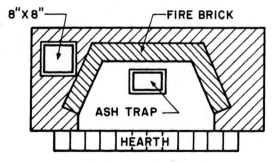

Fig. 6-6 Plan through fireplace

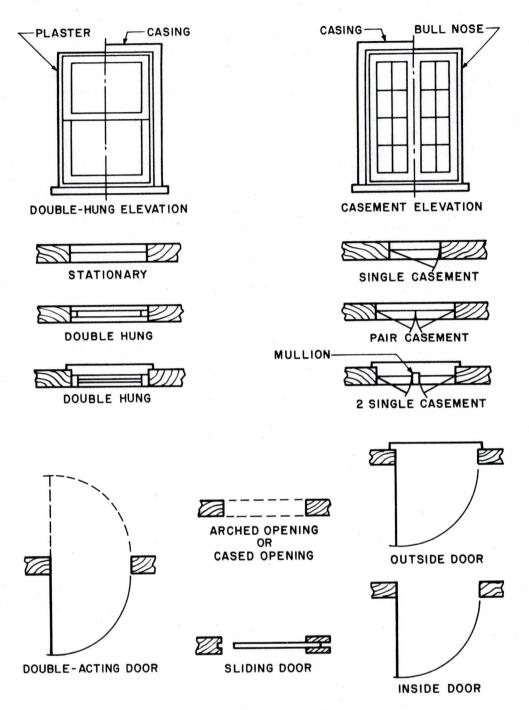

Fig. 6-7 Door and window indicators

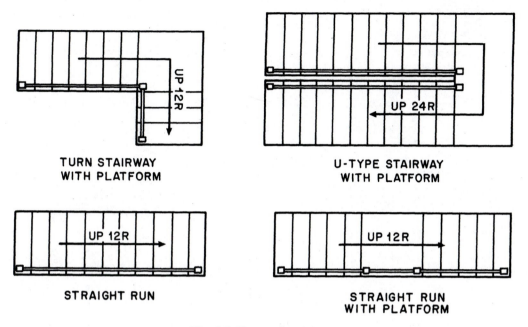

TURN STAIRWAY
WITH PLATFORM

U-TYPE STAIRWAY
WITH PLATFORM

STRAIGHT RUN

STRAIGHT RUN
WITH PLATFORM

Fig. 6-8 Types of stairways

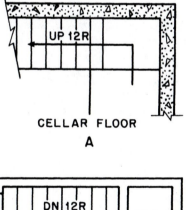

CELLAR FLOOR

A

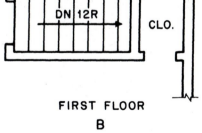

FIRST FLOOR

B

Fig. 6-9 Stairs in plan view

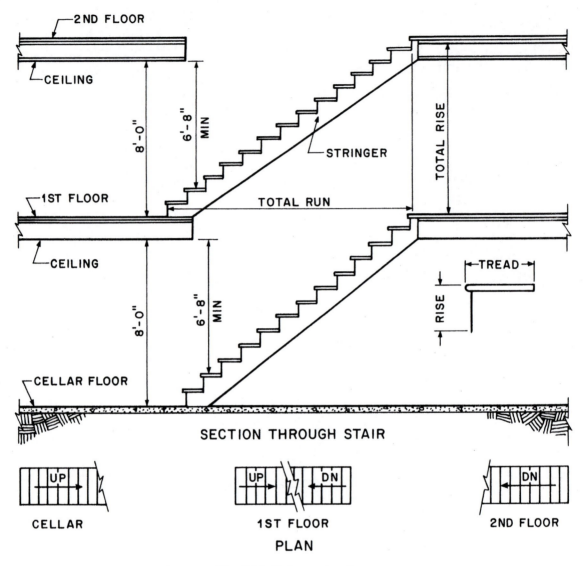

Fig. 6-10 Stair construction

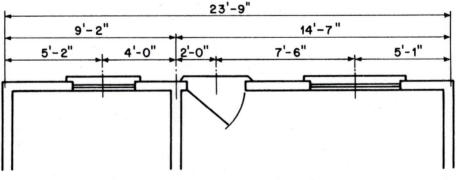

Fig. 6-11 Dimensioning a floor plan

ASSIGNMENT

Make a sketch of the kitchen on Sheet 3/7 in the back of the book. Provide all necessary dimensions. Scale 1/4″ = 1′-0″.

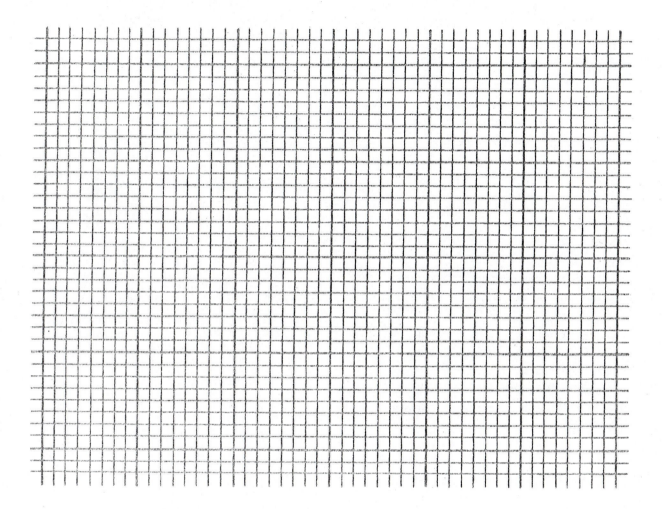

UNIT 7 WALL SECTIONS

OBJECTIVES

- Explain two types of drawings used to illustrate inner construction of buildings.
- List the information that is shown in a typical wall section.

To show the inner construction of a portion of a building, an imaginary cut is made through the elevation. The viewer then imagines that the building is turned, exposing the construction details, figure 7-1.

Wall sections may be either transverse or longitudinal. A *transverse section* is a section taken through the narrowest portion of the building. A *longitudinal section* is a section taken through the longest dimension of the house. Transverse and longitudinal sections are used to show unusual construction details, such as split level construction or stair location, figure 7-2.

BASIC PARTS OF THE WALL SECTION

A typical wall section is usually divided into three basic parts. They are the footing, the sill, and the cornice.

Footing Details

Footings are placed at the base of the foundation wall, figure 7-3. They support the foundation walls and other dead and live loads imposed upon it. The size of the footing and the material used for its construction are listed on the wall section. The material is indicated by symbols. Refer to Unit 1 for an explanation of symbols.

Sill Details

The construction members at the top of the foundation wall comprise the sill detail, figure 7-4. The sill detail shows a portion of the foundation wall, termite shield, sill plate, floor joist, subfloor, finished floor, trim, bottom plate, stud, exterior wall finish, and interior wall finish.

Cornice Details

The cornice detail is similar to the sill detail in that it shows how the various framing members fit together. All of the members are usually sized, called out by note, or denoted by symbol to clarify construction.

There are many different types of cornices used in house construction. Figure 7-5 is a cross-section detail of a snub cornice. This type of cornice is used at the eave line where a moderate projection and ornamentation is required. The width of the plancier board, frieze, and molding may be modified as required.

Figure 7-6 shows a section of a box cornice as it would appear at the gable rake line. Projection of the box cornice varies.

Figure 7-7 shows the type of cornice used where an extended overhang is wanted and the roof is pitched. Notice that the plancier is below the ceiling line due to the roof slope.

The ranch-type cornice is very similar to this type. With a comparatively flat roof, the plancier boards are fastened to the bottom of the roof joists that extend beyond the building as do the lookouts in figure 7-7. This brings the plancier boards on a line with the interior ceiling which is characteristic of the modern ranch-type cornice. The plancier boards are sometimes built up of matched boards or exterior plywood to the required extension.

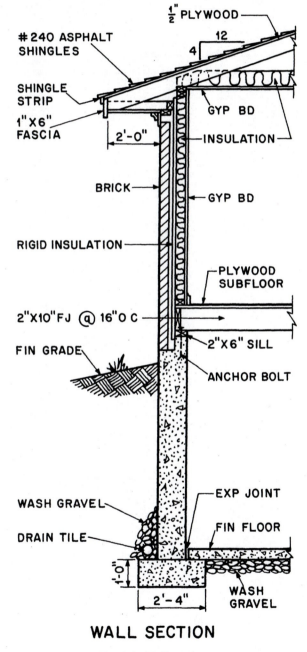

WALL SECTION

Fig. 7-1 Wall section

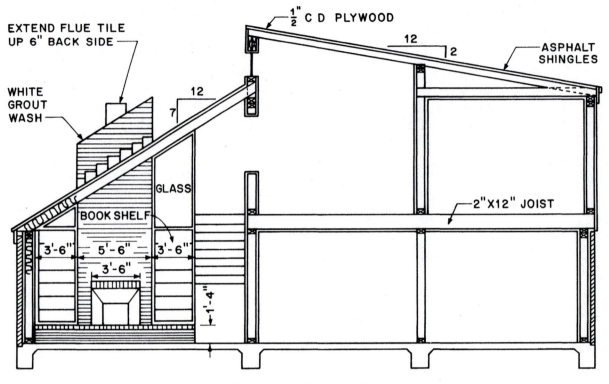

EXTEND FLUE TILE
UP 6" BACK SIDE

WHITE
GROUT
WASH

½" C D PLYWOOD

12
2

ASPHALT
SHINGLES

12
7

GLASS

BOOK SHELF

2"X12" JOIST

3'-6"   5'-6"   3'-6"

3'-6"

1'-4"

Fig. 7-2 Transverse section

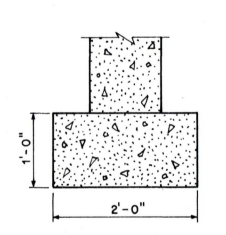

1'-0"

2'-0"

Fig. 7-3 Footing detail

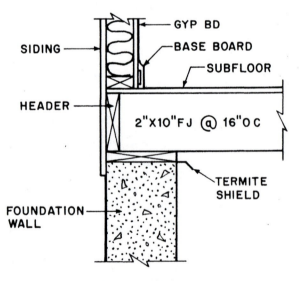

SIDING

HEADER

FOUNDATION
WALL

GYP BD

BASE BOARD

SUBFLOOR

2"X10"FJ @ 16"O C

TERMITE
SHIELD

Fig. 7-4 Sill detail

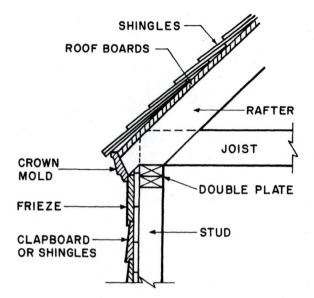

Fig. 7-5 Cross-section detail of snub cornice

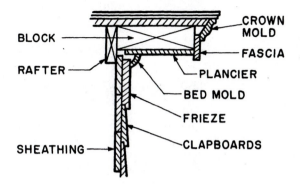
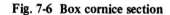

Fig. 7-6 Box cornice section

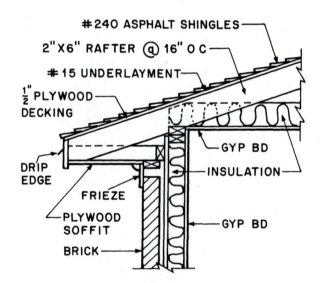

Fig. 7-7 Cornice type used with extended overhang

ASSIGNMENT

A. Answer the following questions, referring to the drawings of the house, sheets 1/7 through 7/7, in the back of the book.

1. What is the size of the floor joists?

2. What size anchor bolts are used?

3. What is the width of the foundation wall?

4. Of what material is the soffit constructed?

5. Where is an expansion joint used?

6. What is the thickness of the sand and gravel fill?

7. What material is used for the subfloor?

8. What is the size of the fascia?

9. What is used as a roof covering?

10. What is the slope of the roof?

11. What is the width of the footings?

12. Where is a vapor barrier used?

13. What is the size of the rafters?

14. What is the size of the ceiling joist?

15. What is placed directly over the roof sheathing?

16. What is nailed on the outside of the studs in Section D-D?

17. Why is copper tubing placed in the stone veneer?

18. Why are wall ties used?

19. What is placed between the studs in Section D-D?

20. What size drainpipe is used?

B. Draw a typical wall section that includes some of the following:
16" x 8" concrete footing, 18" crawl space, concrete block foundation
wall, termite shield, 2" x 12" header, 2" x 12" floor joist, 2" x 4" frame
wall, closed horizontal cornice, plywood soffit, 4" per foot slope, and
asphalt shingles. Draw and call out all of the items included in your wall
section.

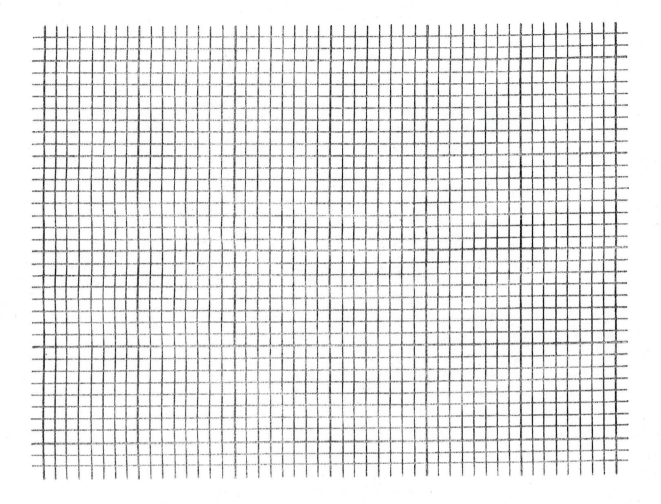

UNIT 8 ELEVATIONS

OBJECTIVES

- Describe the purpose of elevation drawings.
- Explain the features of elevations.

An elevation is an *orthographic drawing* of a building; that is, a drawing based on perpendicular lines or right angles. Elevations show the distinguishing exterior features of a building. They illustrate design principles, materials in symbol and note, and some basic dimensions. Elevations show the object as it is viewed directly in front of each face of the object, figure 8-1. Notice that the elevation appears as a flat surface without depth.

Now observe the front elevation shown in figure 8-2. This front elevation appears the same, whether or not the house has any recess or projection in the front. This front elevation also appears the same whether the house has a garage which projects in front or is flush, figure 8-3. Only by comparing elevations, or by reading the plan, can such information be noted. All of the elevation views must be studied before it is possible to accurately visualize the house.

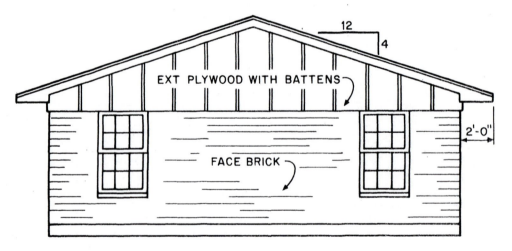

LEFT-SIDE ELEVATION
SCALE $\frac{1}{4}" = 1'-0"$

Fig. 8-1 Left-side elevation

FRONT ELEVATION

Fig. 8-2 Elevations show no depth.

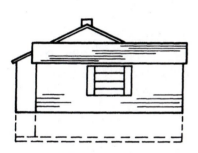

LEFT ELEVATION

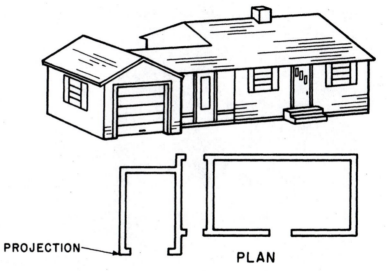

PROJECTION

PLAN

LEFT ELEVATION

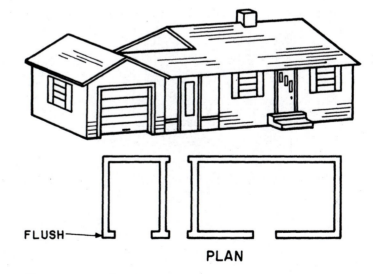

FLUSH

PLAN

Fig. 8-3 Elevations reveal projections.

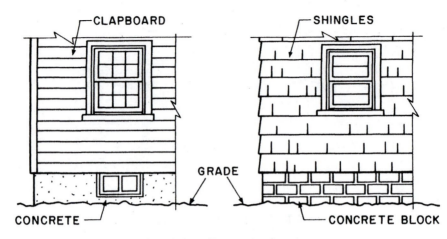

Fig. 8-4 Indicators in elevation

Designating Elevations

There are two methods of designating elevation drawings. In the first method, the architect labels the drawing of the front of the building as the *Front Elevation.* From this view to the right side of the building is called the *Right-Side Elevation;* that on the left, the *Left-Side Elevation.* The drawing of the rear or back of the house is the *Rear Elevation.*

In the second method, the elevations of the building are labeled according to the compass direction in which they face. If the front of the house faces south, this elevation is called the *South Elevation;* the right side is the *East*

Elevation; the left side, the *West Elevation,* and the rear is the *North Elevation.*

Figure 8-4 illustrates a conventional method of showing window frames, sash, and sash light divisions on elevation drawings. Unless special types of frames and glass shapes within the sash are required, glass sizes are not shown.

In most instances, window units are described by specifying the manufacturer's catalog from which the units were selected and the catalog number of the units. Such catalogs give the type and size of frame, sash opening, sash thickness, glass size, and size of the entire

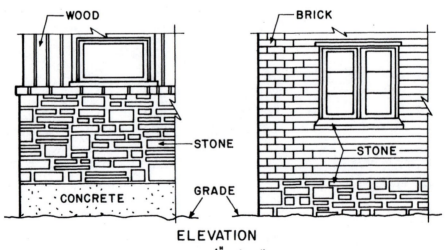

ELEVATION

SCALE $\frac{1}{4}$" = 1'-0"

Fig. 8-5 Symbols used in elevation

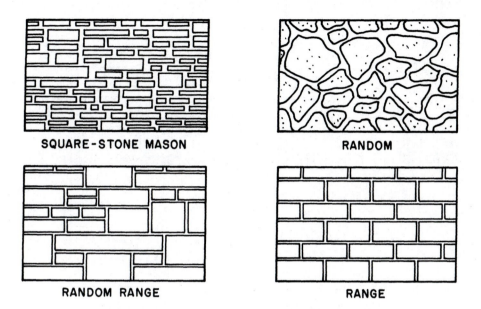

SQUARE-STONE MASON

RANDOM

RANDOM RANGE

RANGE

Fig. 8-6 Elevation symbols for stone and brickwork

opening that should be made in the framed wall to provide proper clearance to house the window frame.

Elevation Symbols and Indicators

The symbols and indicators used in elevation views to show the type, shape, and location of construction are easily understood as they, too, appear in a picturelike representation, figure 8-5. Shingles, stone, and brickwork are shown as they would appear to a person

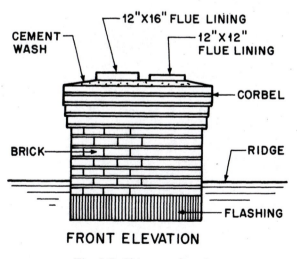

CEMENT WASH

12"X16" FLUE LINING

12"X12" FLUE LINING

CORBEL

BRICK

RIDGE

FLASHING

FRONT ELEVATION

Fig. 8-7 Chimney elevation

looking at the objects themselves. In many cases, the architect further simplifies these symbols by adding a descriptive word or phrase to further define the item.

Elevation symbols not only illustrate the material, but often show peculiarities of construction which cannot be shown in any other view. This is especially true of stone and brickwork, figure 8-6. The chimney in figure 8-7 is built of brick and easily interpreted as such. However, the corbeling shown at the chimney cap is an equally important item and can be observed only in elevation.

Elevation drawings also show footings, foundations, and any other portions of the building which are or will be below finish grade. They are shown using hidden object (dash) lines.

Note that the footings in the elevation in figure 8-8 are shown to be stepped. This is a good method of construction since the building site slopes. The footings will therefore follow the natural contour of the ground.

The roof slope indicated is 4:12. This means that for every rise of 4 feet, there is a run of 12 feet. Roof slope usually is explained in terms of 4 inches rise per 1 foot of run. That

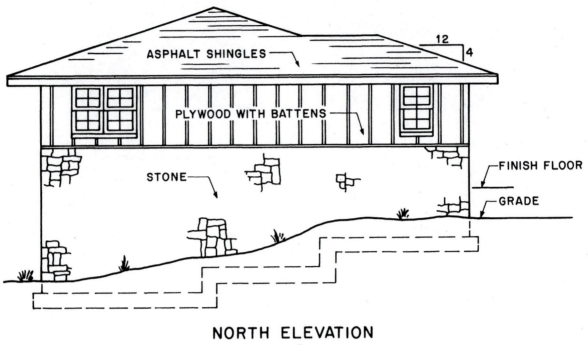

NORTH ELEVATION

SCALE $\frac{1''}{4} = 1' - 0''$

Fig. 8-8 Note the stepped footings

is, for every foot of run, the roof slope rises 4 inches.

Pitch, rise, run, and, *span* are terms used to describe roof layout and construction. Their correct relationships are shown in figure 8-9.

In summary, the following should be noted regarding elevation drawings.

- Elevations have no depth; therefore, all the elevation drawings must be compared so that the house may be correctly visualized.

- Elevations may be designated by the manner in which the sides appear to the viewer, such as Front, Right, Left, Rear; or designated by the compass direction which they face.

- Symbols and indicators are used in elevation views to describe clearly the items of material used or the details of construction.

- Abbreviations are used frequently and the tradesworker must be familiar with their meanings to read the blueprint accurately.

- The specifications must be used in conjunction with the drawings because specifications give construction information which cannot be included on the drawings.

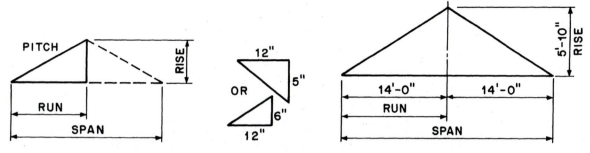

Fig. 8-9 Roof terms and layout

ASSIGNMENT

Answer the following questions, referring to the elevation drawings on Sheets 4/7, 5/7, and 6/7 in the back of the book.

1. What type of siding is used?

2. Of what material are the gutters made?

3. What is the size of the leaders?

4. To what scale are the elevations drawn?

5. What is the distance from the new grade to the finished first floor?

6. What is placed on top of the stone veneer?

7. What is the distance from the finished basement floor to the bottom of the footings?

8. What is the distance from the finished garage floor to the finished first floor?

9. What is the distance from the top of the garage foundation and floor to the finished floor?

10. What is the depth (NW to SE) of the chimney?

11. Of what type of material is the terrace made?

12. What type of flashing is used around the chimney?

13. What two dimensions do elevations have?

14. What are the two methods of labeling elevations?

UNIT 9 SCHEDULES

OBJECTIVES

- Use schedules on blueprints.

- List and explain the symbols used for identification.

A *schedule* is a separate supplemental list which is used to describe windows, doors, and wall finishes. Schedules are not difficult to understand. However, the tradesworker must be able to recognize and interpret them properly.

Usually, a window or door on the plan is designated by a number or letter. In the schedule box, the same letter or number is duplicated with a brief description of the item.

Figure 9-1 illustrates typical door and window schedules and shows their relationship to a floor plan. The arrows from the identifying letter on the plan to the identifying letter in the schedule do not appear on the regular set of plans. This illustration also includes a finish schedule. This schedule briefly describes wall and ceiling finishes, kinds of floors, and similar items.

It must be remembered that schedules are merely short descriptions of the items mentioned. The plan reader must still check the various drawings and the specifications for all necessary information.

Schedules aid greatly in estimating cost of items. For example, by studying the floor plan and schedule in figure 9-1, it is apparent that three Type 2 windows are required. Similarly, it can be determined that wood baseboard is required in the living and dining room areas only. The specifications would state the size lumber to be used and the size and type of shoe or base mold, if any. Thus, it is possible to quickly estimate the total base required; then, its cost can be determined.

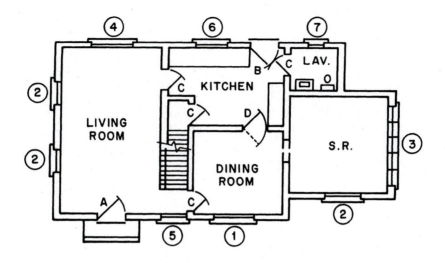

WINDOW SCHEDULE		
Type	Size	Remarks
1	2'-0" x 4'-6"	Double Hung
2	2'-4" x 4'-6"	Double Hung
3	1'-8" x 4'-6"	Casement
4	2'-2" x 3'-6"	Double Hung
5	2'-2" x 3'-0"	Fixed Sash
6	3'-0" x 3'-2"	Double Hung
7	2'-0" x 2'-2"	Casement

DOOR SCHEDULE		
Type	Size	Remarks
A	3'-0" x 7'-0" x 1 3/4"	4 Pnl. 2 Lights
B	2'-8" x 6'-10" x 1 3/4"	1 Pnl. 4 Lights
C	2'-6" x 6'-8" x 1 3/8"	6 Panel
D	2'-8" x 6'-8" x 1 3/8"	6 Panel

ROOM	FLOOR	WALLS	CEILINGS	BASE	WAINSCOTT
L	Oak	Gyp Bd	Gyp Bd	Wood	
D	Oak	Gyp Bd	Gyp Bd	Wood	Pine Panel'G
K	Asphalt	Gyp Bd	Gyp Bd	Asp. Tile	
S.R.	Vinyl	Birch Ply	Gyp Bd		
Lav.	Cer. Tile	Cer. Tile	Gyp Bd	Cer. Tile	Cer. Tile

Fig. 9-1 Door and window schedules

WINDOW SCHEDULE		
Type	Size	Remarks
100	5'-0" x 4'-6"	Fixed 24L.
101	2'-4" x 4'-6"	D.H.
102	2'-6" x 4'-6"	D.H.
103	2'-8" x 3'-0"	D.H.
104	2'-2" x 3'-0"	Casement
105	1'-6" x 4'-6"	D.H.

DOOR SCHEDULE		
Type	Size	Remarks
A	3'-0" x 7'-0" x 1 3/4"	6 Pnl. Col.
B	2'-10" x 6'-10" x 1 3/4"	3 Pnl. Gl.
C	2'-6" x 6'-8" x 1 3/8"	6 Pnl.
D	2'-4" x 6'-8" x 1 3/8"	6 Pnl.
E	2'-0" x 4'-0" x 1 1/8"	Pine. Bat.
F	2'-8" x 6'-8" x 1 3/8"	Doub. Acy.

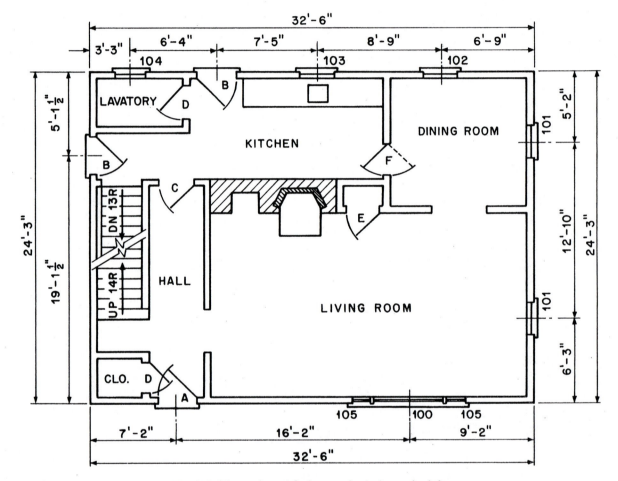

Fig. 9-2 Floor plan with door and window schedules

ASSIGNMENT

A.   Answer the following questions, referring to the floor plan with the door and window schedule in figure 9-2.

1. How many windows are there in the dining room? *2*

2. What is the size of window 100? *5' x 4'-6"*

3. Can window 100 be used for ventilation? *no*

4. How many panels are there in a B door? *3*

5. How many B doors are there? *2*

6. What is the width of a 105 window? *1'-6"*

7. What is the width of a 103 window? *2'-8"*

8. How many lights of glass are there in the C doors? *6*

9. What is the stile of a door?

10. What is a batten or battened door? *X*

11. What is the dimension from the center of the front door to the left front corner of the house?

12. How many windows are there in the left wall?

13. How many C doors are required?

14. Stock sash are usually of what thickness?

15. What is meant by D.H.?

16. How many casement windows are shown?

17. What is the dimension from the center of the kitchen window to the center of the dining room window?

18. Interpret the indicator $\frac{6}{1}$ as it applies to windows. *top sash has six panes of glass*

1 on bottom

19. What is the dimension from the center of the living room window to the center of the dining room window?

20. What is the area of the two sides of an A door? 42 ft²

21. The sash fasteners for the D.H. windows cost $1.10 per unit. What is the total cost of all the fasteners needed?

22. What is the difference in width between an A door and a D door? 8"

23. How many cased openings are shown?

24. What is meant by a double-acting door?

25. Does the casement window swing in or out?

26. The center of A door is how far from the front right corner as viewed from the front of the house?

27. How many outside doors are there in the kitchen? 2

28. How many different sizes of doors are there in the hall? 3

29. How many doors are there in the living room?

30. What is the thickness of an inside door?

31. What is the standard thickness of an outside door?

32. How many different heights of windows are there? 2

33. List the various heights of the windows. 4'-6"/3'-0"

34. Give the area of a 100 window. 22.5

35. What is the width of the lavatory window?

36. If the tops of all the sash were 6'-6" from the floor line, how high would the bottom of the 105 window be above the floor line? 2'

37. How many closet doors are shown? 2

38. How many different size doors are there in the kitchen?

39. What is the dimension from the center of the dining room window in the rear wall to the rear left corner?

40. How many window units are there in the house?

41. What is the lock rail of a door?

42. How many doors are there in the outside walls of the house?

43. How many rough openings are needed in the front wall?

44. What is the dimension from the center of a 100 window to the front left corner of the house?

45. How many different kinds of windows are there in this house? List them.

46. What is the dimension from the center of 104 window to the center of 102 window?

B. Prepare a finish schedule for the floor plan and schedule in figure 9-2.

UNIT 10   THE FLOOR FRAME PLAN

OBJECTIVES

- State the purpose of framing plans.
- Recognize the location and terms of various members of floor framing.
- Explain and utilize the various methods of framing openings in floor platforms.

A *floor framing plan* shows the location and size of the various framing members in the floor frame, figure 10-1.

In many cases, a floor frame plan is not included in a typical set of drawings. In these instances, the actual framing is the responsibility

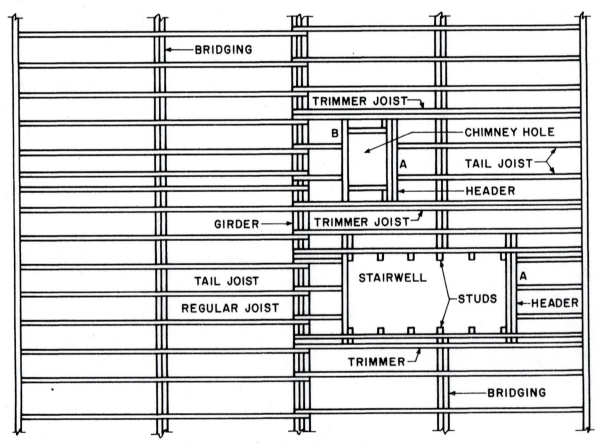

Fig. 10-1 Floor framing plan

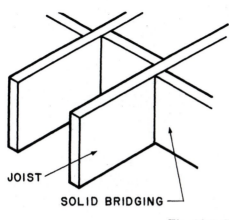

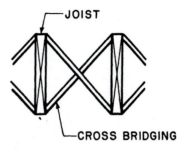

Fig. 10-2 Solid and cross bridging

of the contractor. Framing methods and techniques are also omitted in most specifications because it is difficult and lengthy to describe the work clearly. The carpenter, therefore, must know some of the basic principles of typical floor framing.

Floor framing plans provide a clear picture of the framed structure. They are of great value to the estimator and to the contractors of the various trades involved in the construction of the building. In many cases, local building codes explain the required size of joists for various spans, required headers to carry joists to well holes and bearing partitions, and the minimum requirements of the floor members.

Floor Joists

Regular joists are framing members that are continuous from bearing to bearing and carry a distributed floor load. *Headers* are joists that are placed at right angles to the floor joist. Headers are usually doubled if they support four or more joists, unless the header is close to a joist bearing. In this case, a single header may be used, figure 10-1, note A and B.

Trimmers are floor joists that are used to trim an opening to the required size. They are also used to support headers. They may be either regular joists or extra joists parallel to the regular joist. If the trimmer joists are supported by studs, a single trimmer can be used.

If they are not supported over a long span, they should be doubled to support the header joist.

Tail joists are shortened regular floor joists that extend from the headers to the supporting girder or sidewall bearing.

Bridging

To help distribute the floor load over the floor frame, either solid or cross bridging is used, figure 10-2. Diagonal bracing is placed from the top of one joist to the bottom of an adjacent one. In most cases, they are either 1 x 4 or fabricated steel straps. Note: A 1" x 4" is a piece of lumber that is actually 3/4 inch thick and 3 1/2 inches wide. Figure 10-3 gives gives other nominal and actual (dressed) lumber measurements.

Solid bridging is cut from the same size stock as the floor joists. Once cut they are placed between the joists. They may be placed in a straight line or offset for easier installation. In most cases, the bridging is placed in a continuous line in the center of the floor span.

Girders

A *girder* is a horizontal structural member that supports the ends of floor joists, figure 10-4. Girders are usually constructed from solid wood, laminated wood, or steel of various shapes.

THICKNESS		WIDTH	
Nominal	Dressed	Nominal	Dressed
1	3/4	2	1 1/2
1 1/4	1	3	2 1/2
1 1/2	1 1/4	4	3 1/2
2	1 1/2	5	4 1/2
2 1/2	2	6	5 1/2
3	2 1/2	7	6 1/2
3 1/2	3	8	7 1/4
4	3 1/2	9	8 1/4
		10	9 1/4
		11	10 1/4
		12	11 1/4

Note: With nominal sizes, the inch measurement is often assumed. For instance, a 2″ X 4″ piece of lumber is simply called a 2 X 4 (two-by-four).

Fig. 10-3 Nominal and dressed lumber sizes

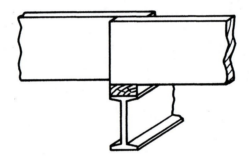

Fig. 10-4 Steel girder

A girder is usually placed near the center of the structure. It replaces a load-bearing frame wall. It is usually supported by the foundation wall and columns.

ASSIGNMENT

A. Study the information shown in the floor plan of the house on Sheet 3/7 in the back of the book. Draw a floor framing plan of the living room and dining area. Show all required floor structural members and their names. Scale 1/4″ = 1′-0″.

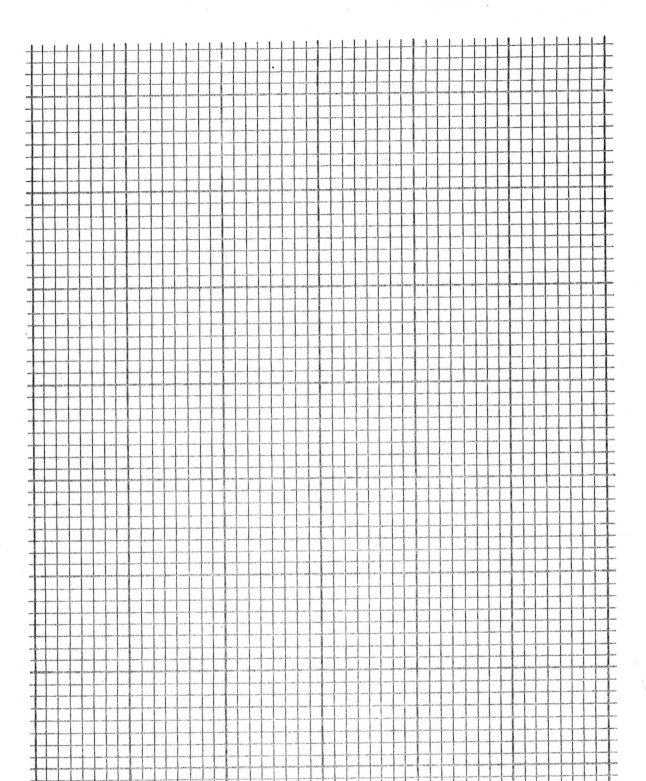

B.   Draw the floor framing plan of the kitchen and breakfast area on Sheet 3/7 in the back of the book. Scale 1/4″ = 1′-0″.

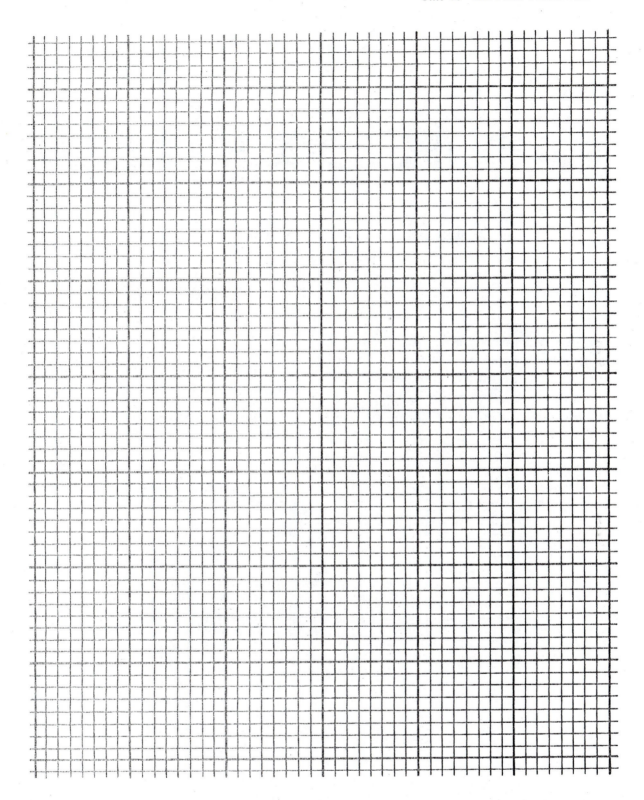

UNIT 11   THE ROOF FRAME PLAN

OBJECTIVES

- List and identify structural members of a roof frame plan.
- Explain the principles of framing openings in a roof frame plan.
- Sketch a roof frame plan.

The roof frame is built of structural members that are designed, located, and assembled to carry the distributed and concentrated loads placed upon them. The roof frame plan, figure 11-1, shows the location and size of each individual member.

Roof Framing Members

The following are roof framing members that are commonly used in construction.

- **Common rafter** is a rafter that extends from the bearing plate to the ridge.

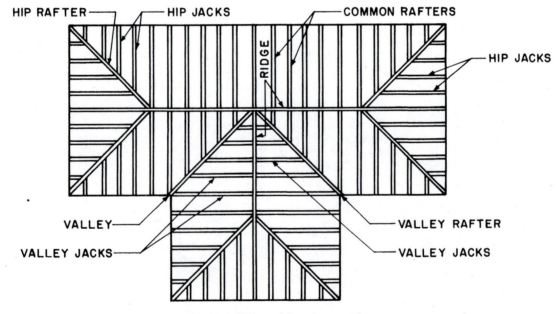

Fig. 11-1 Hip roof framing members

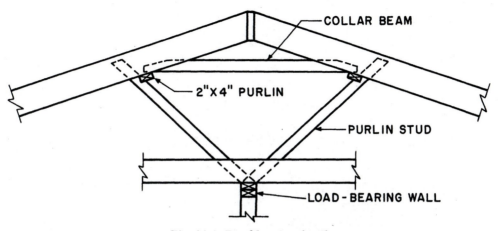

COLLAR BEAM

2"X 4" PURLIN

PURLIN STUD

LOAD - BEARING WALL

Fig. 11-2 Roof bracing detail

Common rafters support the sheathing and roof covering.

- **Valley rafter** is a rafter that extends from an interior corner to a ridge. The lowest portion of the valley rafter rests on the bearing plate.

- **Jack rafter** is a rafter that extends from the ridge to a valley rafter or from a bearing plate to a hip rafter.

- **Hip rafter** is a rafter that extends from an outside corner to a ridge. It connects the intersecting slopes of the hip.

- **Cripple rafter** is a short rafter that extends from a hip to a valley, but does not extend to a ridge.

- **Collar beam** is a bracing member that is. placed between two opposite rafters.

- **Ridgeboard** is a horizontal framing member that connects the uppermost portions of the rafters.

- **Bearing plate** is a horizontal framing member that is fastened to the top of a frame wall.

- **Purlin** is a horizontal roofing framing member (usually a 2 x 4) that is nailed to the underside of the rafters, figure 11-2.

- **Purlin stud** is a structural member attached to a rafter and a load-bearing wall.

It is used to brace the roof against live and dead loads.

Special Roof Openings

Openings in roofs, such as a chimney hole or scuttle hole, are shown in the framing plan in the same manner as the stairwell is shown in a floor plan, figure 11-3.

The framing plan for a gable or hip dormer is similar in part to the framing plan for the gable and hip roof plan.

Roof Trusses

Roof trusses are now used extensively in residential construction. They are easy to erect and have great strength, thereby saving time and money.

Trusses are engineered framing members that are usually fabricated in a shop under controlled conditions. They are built in special jigs that tightly clamp them until the connector plates have been properly located and fastened to the cords and webs.

Trusses are available in many different sizes and styles, but the most popular style is the fink truss, figure 11-4. A *fink truss* is constructed of two top chords, one bottom chord, and webs that tie all the chords together. The webs and chords are usually connected by metal or plywood gusset plates.

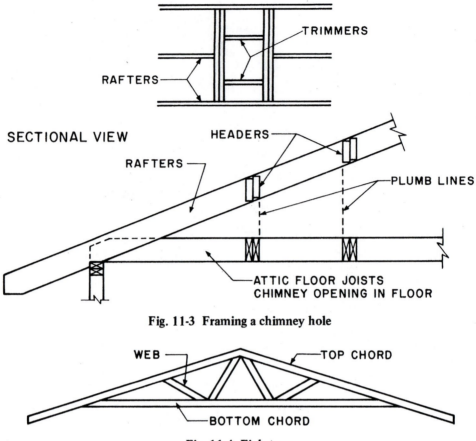

SECTIONAL VIEW

Fig. 11-3 Framing a chimney hole

Fig. 11-4 Fink truss

ASSIGNMENT

A.   Fill in the blanks.

1.   A _____ rafter is a rafter that extends from the bearing plate to the ridge.

2.   A _____ plate is a horizontal framing member that is fastened to the top of a frame wall.

3.   The most popular truss is a _____.

4.   The webs and chords are usually connected by _____.

5.   A _____ is a horizontal roof framing member that is nailed to the underside of the rafters.

6.   A _____ is a horizontal framing member that connects the uppermost portions of the rafters.

7.   A _____ is a bracing member that is placed between two opposite rafters.

8.   A _____ is a rafter that extends from an outside corner to a ridge.

B.   Sketch and identify the structural members of a roof frame plan.

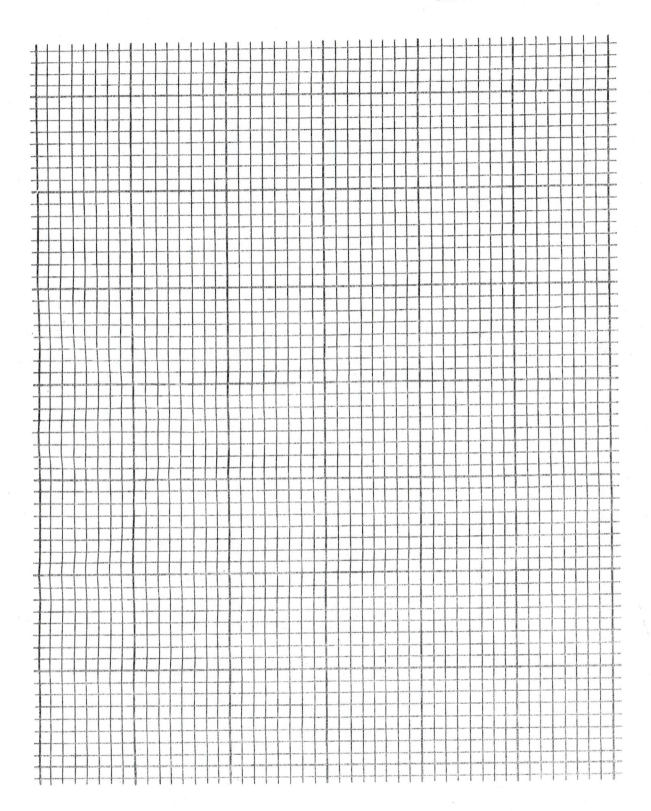

UNIT 12 DETAILS-WINDOWS AND DOORS

OBJECTIVES

- Read window and door details.
- Make a rough sketch of the detail and assembly of window and door units in various types of wall construction.
- List types and descriptions of window and door units.

Types, styles, and methods of fabricating window and door frames have become so varied that manufacturers provide catalogs that list the various standard units produced by their company.

It is always good practice to study catalogs carefully before making a selection. It is also necessary to know some of the basic information which applies to all window and door units, regardless of style. This is important to the plan reader because, in many instances, both special types and standard types are used by the architect in preparing drawings.

WINDOW DETAILS

The location of the window opening is usually given on the plans by showing the centerline of the opening. The rough opening is determined from this centerline. For construction and installation of the window, however, details quite often must be followed. The parts of a window are shown in figure 12-1.

The following is a list of the basic types of windows. Typical catalog information as supplied by manufacturers is used in the descriptions.

Double-Hung Windows

Double-hung windows are the most commonly used windows. They are available in wood, aluminum, or steel. Weights and balances are often eliminated in the newer designs by such devices as concealed pressure strips which force the sash against the parting stop.

Table 12-1 illustrates the variety, types and sizes of double-hung windows available from the standard stock sizes.

To figure openings for multiple units:

- Add sash opening widths plus 2 3/16 inches for each mullion.
- For rough stud openings, add 3 1/2 inches to overall sash opening width.
- For masonry openings, add 4 inches to overall sash opening width.

Figure 12-2 shows the installation of double-hung windows. The first shows how it is installed in a frame wall with reversible stops in correct position for 3/4-inch sheathing. For 1/2-inch sheathing, the position is reversed. The next two show the installation in a brick veneer wall and in a block wall.

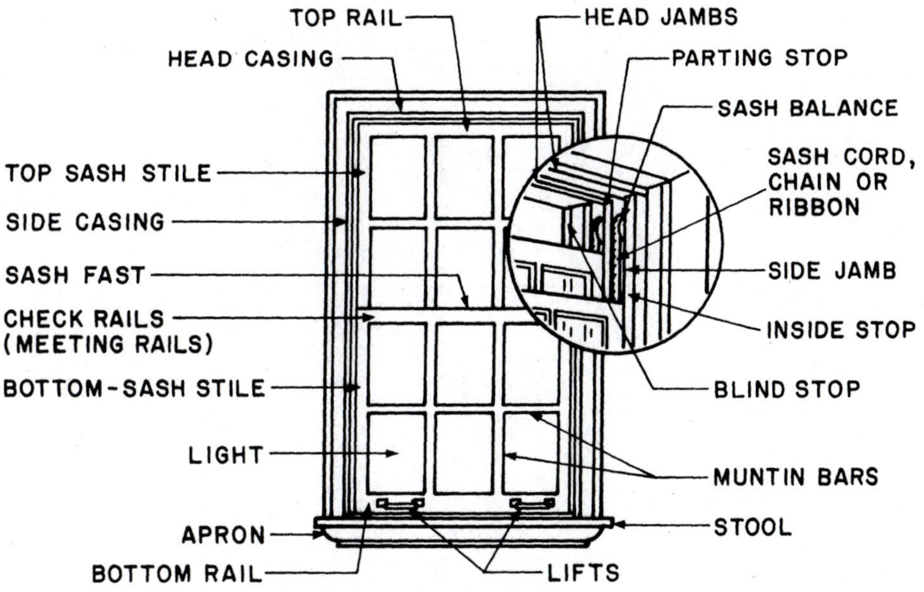

Fig. 12-1 Parts of a window

Glass sizes shown for Types 61 and 81 are for bottom sash only. SEE TYPES 66 and 88 for glass sizes in top sash.

WINDOW TYPES			11	22	66	88	61	81	23	69
SASH OPENING	ROUGH STUD OPENING	MASONRY OPENING	GLASS SIZE	GLASS SIZE	GLASS SIZE	GLASS SIZE	GLASS SIZE	GLASS SIZE	GLASS SIZE	GLASS SIZE
2-0x3-2	2-3½x3-5½	2-4x3-6¾	20½x16¼	20½x8	6⅝x8		20½x16¼			
3-10	4-1½	4-2¾	20¼	10	10		20¼			
4-2	4-5½	4-6¾	22¼	11	11		22¼			
4-6	4-9½	4-10¾	24¼	12	12		24¼			
5-2	5-5½	5-6¾	28¼	14	14		28¼			
5-6	5-9½	5-10¾							20½x12	6⅝x12
2-4x3-2	2-7½x3-5½	2-8x3-6¾	24½x16¼	24½x8	8x8		24½x16¼			
3-10	4-1½	4-2¾	20¼	10	10		20¼			
4-2	4-5½	4-6¾	22¼	11	11		22¼			
4-6	4-9½	4-10¾	24¼	12	12		24¼			
5-2	5-5½	5-6¾	28¼	14	14		28¼			
5-6	5-9½	5-10¾							24½x12	8x12
2-8x3-2	2-11½x3-5½	3-0x3-6¾	28½x16¼	28½x8	9 5/16x8		28½x16¼			
3-10	4-1½	4-2¾	20¼	10	10		20¼			
4-2	4-5½	4-6¾	22¼	11	11		22¼			
4-6	4-9½	4-10¾	24¼	12	12		24¼			
5-2	5-5½	5-6¾	28¼	14	14		28¼			
5-6	5-9½	5-10¾							28½x12	9 5/16x12
3-0x3-2	3-3½x3-5½	3-4x3-6¾	32½x16¼	32½x8		7 15/16x8		32½x16¼		
3-10	4-1½	4-2¾	20¼	10		10		20¼		
4-2	4-5½	4-6¾	22¼	11		11		22¼		
4-6	4-9½	4-10¾	24¼	12		12		24¼		
5-2	5-5½	5-6¾	28¼	14		14		28¼		
5-6	5-9½	5-10¾							32½x12	
3-4x3-2	3-7½x3-5½	3-8x3-6¾	36½x16¼	36½x8		8 15/16x8		36½x16¼		
3-10	4-1½	4-2¾	20¼	10		10		20¼		
4-2	4-5½	4-6¾	22¼	11		11		22¼		
4-6	4-9½	4-10¾	24¼	12		12		24¼		
5-2	5-5½	5-6¾	28¼	14		14		28¼		
5-6	5-9½	5-10¾							36½x12	

Table 12-1 Andersen double-hung wood window units *(Courtesy of Andersen Corp., Bayport, Michigan)*

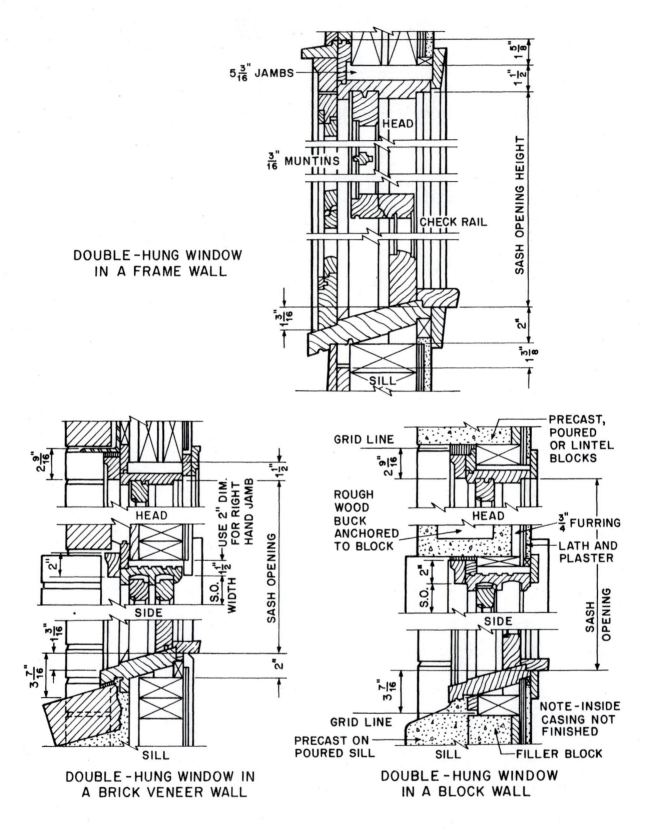

Fig. 12-2 Construction and installation of double-hung windows

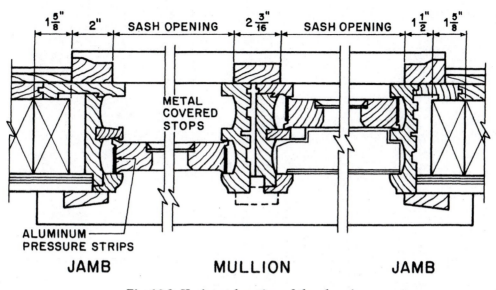

Fig. 12-3 Horizontal section of the plan view

Figure 12-3 illustrates the horizontal section of the plan view. Figure 12-4 shows the corner construction using standing 2-inch casing and regular sill horns.

Casement Windows

The most widely used casement is the outswinging wood type. Figure 12-5 shows construction and installation details of the Woodco wood casement windows. Such details are typical of those furnished by each manufacturer. But, as manufacturers differ slightly in construction or installation of their windows, care must be taken to study the particular catalog for the window specified.

Manufacturers supply information including masonry opening, study opening, and sash opening dimensions for the variety of casement window sizes they make, figure 12-6. The type of jamb desired, regular lath or plaster construction or dry wall construction, must be specified. The jambs are generally 5 1/4 inches or 5 3/8 inches for regular construction; 5 inches for dry wall.

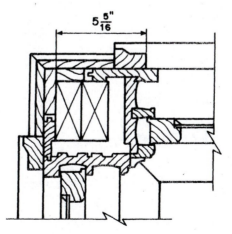

Fig. 12-4 Corner construction

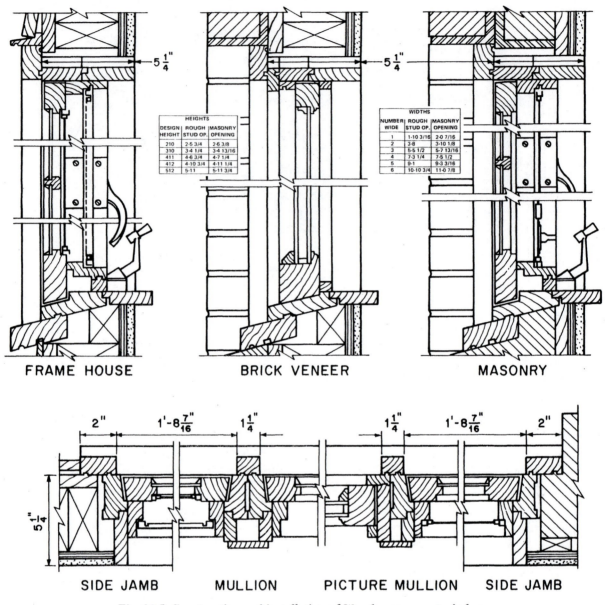

HEIGHTS		
DESIGN HEIGHT	ROUGH STUD OP.	MASONRY OPENING
210	2-5 3/4	2-6 3/8
310	3-4 1/4	3-4 13/16
411	4-6 3/4	4-7 1/4
412	4-10 3/4	4-11 1/4
512	5-11	5-11 3/4

WIDTHS		
NUMBER WIDE	ROUGH STUD OP.	MASONRY OPENING
1	1-10 3/16	2-0 7/16
2	3-8	3-10 1/8
3	5-5 1/2	5-7 13/16
4	7-3 1/4	7-5 1/2
5	9-1	9-3 3/16
6	10-10 3/4	11-0 7/8

FRAME HOUSE   BRICK VENEER   MASONRY

SIDE JAMB   MULLION   PICTURE MULLION   SIDE JAMB

Fig. 12-5 Construction and installation of Woodco casement windows

Sliding or Gliding Windows

The sliding type window is one which is growing in popularity and use. The details of the Woodco Slider® in figure 12-7 are taken from the Woodco catalog.

Multiple Wood Unit Window Combinations

Multiple combinations of window units are often used in groups, ribbons, picture window combinations, and stacks. The follow-

ing information on Flexivent® units is taken from the Andersen catalog.

The installation and construction details are shown in figure 12-8. While there is no difference between the wood sash in a frame building and the wood sash in a masonry building, the window frame is usually placed after an opening has been framed for it. In the masonry building, the frame is set up and plumbed in position and the masonry is built around it.

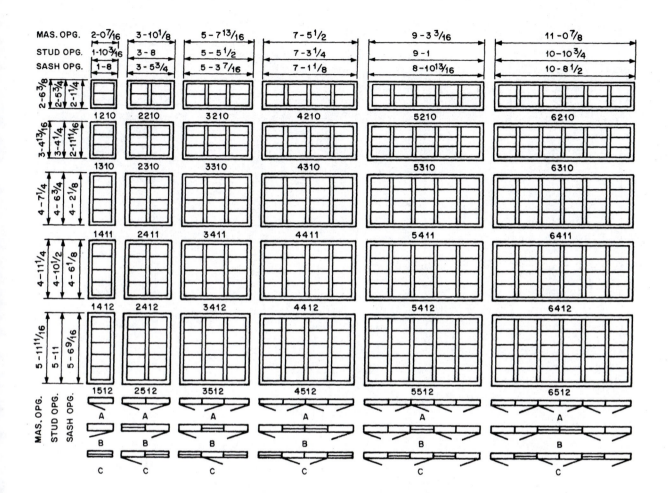

PICTURE WINDOW UNITS

$\frac{1}{4}''$ PLATE GLASS THERMOPANE $\frac{3}{16}''$ A QUALITY GLASS

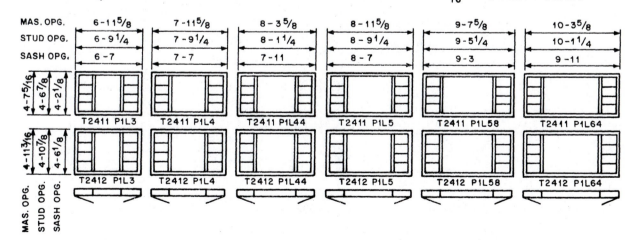

Fig. 12-6 Sizes of Woodco casement windows

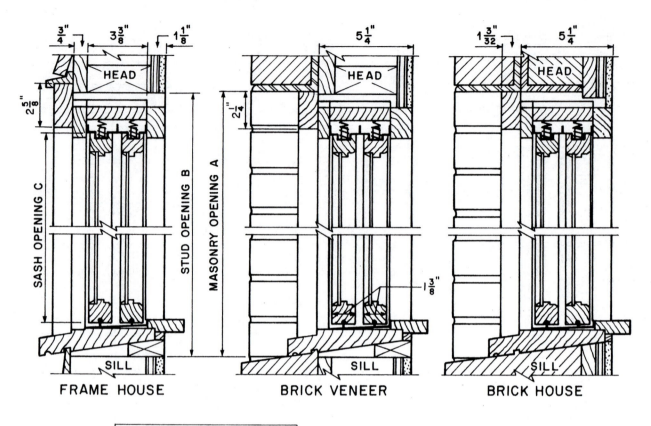

FRAME HOUSE   BRICK VENEER   BRICK HOUSE

WIDTHS			
DESIGN	A	B	C
16	40 7/16	38 1/2	36 7/16
20	48 7/16	46 1/2	44 7/16
24	56 7/16	54 1/2	52 7/16
28	64 7/16	62 1/2	60 7/16

HEIGHTS							
DESIGN	A	B	C	DESIGN	A	B	C
20	26 3/4	26 5/8	22 3/8	36	42 3/4	42 5/8	38 3/8
26	32 3/4	32 5/8	28 3/8	44	50 3/4	50 5/8	46 3/8
32	38 3/4	38 5/8	34 3/8	48	54 3/4	54 5/8	50 3/8

DESIGN NUMBER	SASH SIZE	ROUGH STUD OPENING
16 - 20	3-0 x 2-0	38 1/2 x 26 5/8
16 - 26	2-6	32 5/8
16 - 32	3-0	38 5/8
16 - 36	3-4	42 5/8
16 - 44	4-0	50 5/8
20 - 20	3-8 x 2-0	46 1/2 x 26 5/8
20 - 26	2-6	32 5/8
20 - 32	3-0	38 5/8
20 - 36	3-4	42 5/8
20 - 44	4-0	50 5/8
24 - 20	4-4 x 2-0	54 1/2 x 26 5/8
24 - 26	2-6	32 5/8
24 - 32	3-0	38 5/8
24 - 36	3-4	42 5/8
24 - 44	4-0	50 5/8
28 - 20	5-0 x 2-0	62 1/2 x 26 5/8
28 - 26	2-6	32 5/8
38 - 32	3-0	38 5/8
28 - 36	3-4	42 5/8
28 - 44	4-0	50 5/8
28 - 48	4-4	54 5/8

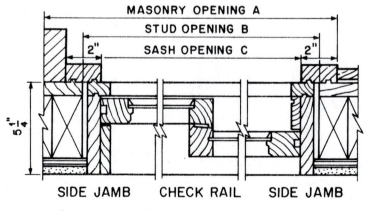

SIDE JAMB   CHECK RAIL   SIDE JAMB

Fig. 12-7 Construction and installation of Woodco Slider® windows

OPENING SIZES
FOR MULTIPLE UNIT COMBINATIONS

Overall sash opening width — The sum of the individual sash openings plus 3″ for each mullion section.

Overall rough stud opening width — Add 3 1/2″ to overall sash opening width.

Overall masonry opening width — (2″ Brick Molding) — Add 6″ to overall sash opening width.

Overall sash opening heights — The sum of the individual sash opening heights plus 2 7/8″ for each stack section.

Overall rough stud opening height — If 1 1/8″ x 4″ subsill is used, add 5 1/4″ to overall sash opening height. If no subsill is used, add 3 5/8″ to overall sash opening height.

Overall masonry opening height — (2″ Brick Molding and 1 1/8″ subsill) — Add 5 3/4″ to overall sash opening height.

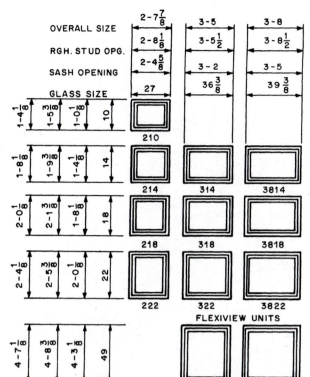

FLEXIVIEW UNITS

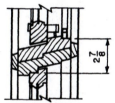

SECTION A-A
STACK SECTION WITH AWNING VENT SASH — ALSO CASEMENT MULLION.

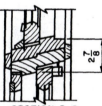

SECTION B-B
STACK SECTION WITH UPPER SASH FIXED AND LOWER SASH HOPPER VENT.

SECTION C-C
MULLION WITH AWNING VENT SASH.

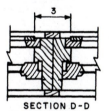

SECTION D-D
MULLION WITH HOPPER VENT SASH.

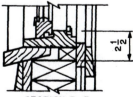

SECTION E-E
SILL SECTION WHEN INSTALLED AS A CASEMENT.

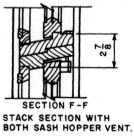

SECTION F-F
STACK SECTION WITH BOTH SASH HOPPER VENT.

TYPICAL ARRANGEMENTS
FLEXIVENT UNIT NO. 318

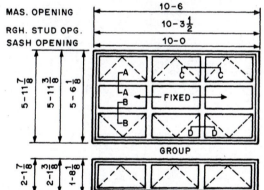

GROUP

RIBBON

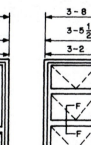

PICTURE WINDOW   STACK

Fig. 12-8 Multiple unit combinations

Although both box and rough-type frames are used in masonry walls, the steel frame and sash is perhaps the most commonly used, particularly in commercial construction.

Even in residential masonry work, metal frames and sash are widely used. Usually the manufacturer of the sash provides working drawings showing the proper installation of the product.

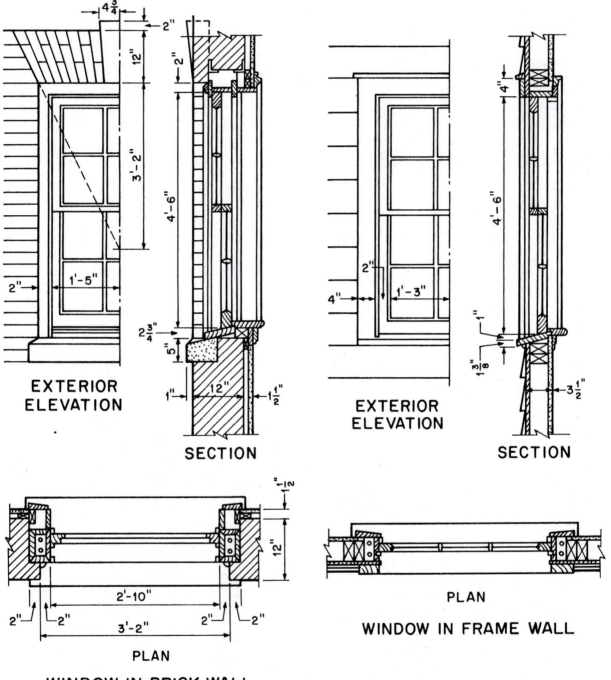

Fig. 12-9 Windows in masonry and frame construction

The mechanic must be able to interpret these details correctly.

WINDOW FRAME MEASUREMENTS

Manufacturers' catalogs give detailed information on how to order specific stock windows. When ordering windows custom-made from a mill, however, it is necessary to supply the mill with accurate information in the proper sequence. To note window frame measurements properly, the horizontal measurements are given first, the vertical measurements second, and the thickness of the wall third.

In figure 12-9, the masonry wall opening is shown as 3'-2". The frame opening, which is the inside distance between the side jambs, is given as 2'-10". The thickness of the masonry wall is 12 inches as shown. The vertical frame opening is 4'-6" and is the same as the vertical sash opening. The vertical masonry opening would be 4'-10 3/4" (2 3/4" + 4'-6" + 2").

To describe the sash, the measurements are given in the sequence described above. In this case, the sash is stated as follows:

Double-hung sash (D.H.) opening
2'-10" x 4'-6" sash
6 lights over 6 lights $\left(\dfrac{6}{6}\right)$ — 1 3/8" thick.

In some cases where the lights of glass in each sash are of a different number, size, or shape, the approximate size of the glass and number of lights in each sash is given.

In figure 12-9, the width of the window jamb for the window frame in the masonry wall is made up of a frame jamb and a false jamb. These jambs are made in widths to accommodate the thickness of the masonry wall and plaster. The jambs of the frame in a frame wall are generally 5 3/8 inches for plastered walls and 5 inches for dry wall construction.

DOOR DETAILS

Like windows, doors are available in a wide variety of styles and sizes for both exterior and interior uses, figure 12-10. Residential exterior doors are generally 1 3/4 inch in thickness and interior doors 1 3/8 inch.

Illustrated in Table 12-2 are a few Curtis door frames for exterior use. C-1980 provides

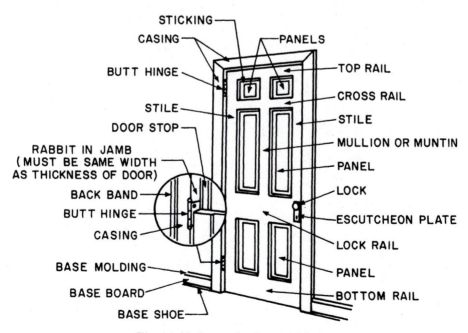

Fig. 12-10 Parts of a door and frame

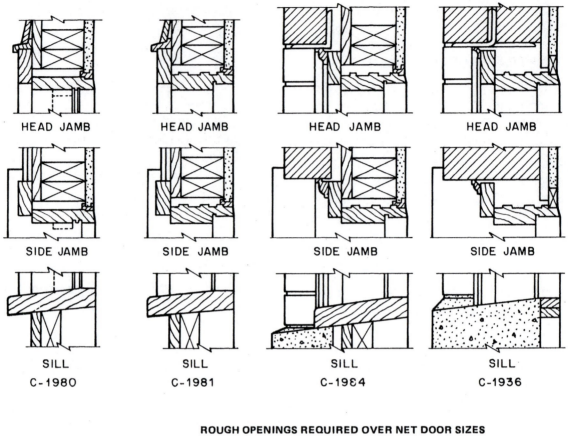

ROUGH OPENINGS REQUIRED OVER NET DOOR SIZES				
	C-1980	**C-1981**	**C-1984**	**C-1986**
In Width	3 1/2''	3 1/4''	3 1/4''	6 1/8''
In Height.	4 1/4''	4 1/8''	4 1/8''	3 5/8''

Rough opening heights for masonry wall frames are taken from the finished floor.

Table 12-2 Curtis door frames for exterior use

for planted stops. Such a frame is designed for an outswinging door. Where an inswinging door is required, the side and head jambs are provided with a rabbetted jamb, C-1981. C-1984 is designed for brick veneer construction and includes a sill. C-1986 is used for masonry construction and is provided without a sill.

Interior door frames, figure 12-11, consist of a head and two side jambs. They form the height and width of the milled door to be used in the opening. The door casings and the stops are added to complete the opening.

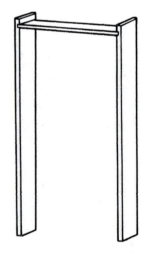

Fig. 12-11 Interior door frame

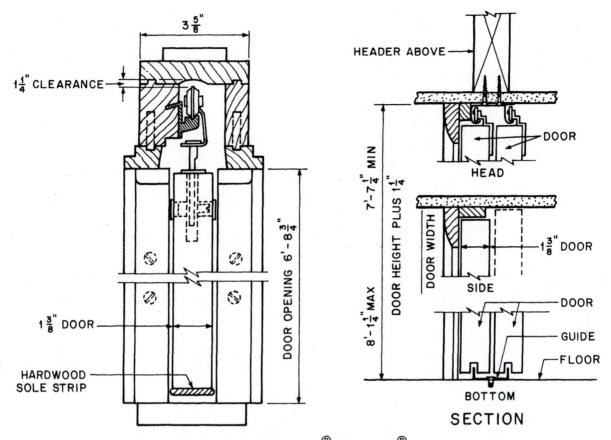

Fig. 12-12 Curtis Gliding® and By-Pass® units

Special types of interior doors include sliding or gliding doors, accordion or folding doors, and glass doors. These are installed according to the specific details provided by the manufacturer. Illustrated in figure 12-12 are details of the Curtis Gliding® and By-Pass® units.

There are instances, especially when ordering hardware, when the hand of a door must be identified. If the butts or hinges of the door are on the left as you stand in front of it and the door swings away from you, it is a left-hand door. If it is on the right, it is a right-hand door.

ASSIGNMENT

A. Answer the following questions, referring to the Andersen double-hung windows, Table 12-1 and figures 12-1, 12-2, and 12-3.

1. If the sash opening is 2'-8" x 4'-6", what is the glass size for the bottom sash, type 61?

2. With the same sash opening as in #1 above, what is the glass size for the top sash, type 61?

3. What must the rough stud opening be if the sash opening is 3'-4" x 5'-2"?

4. What is the masonry opening required for a sash opening of 2'-4" x 5'-2"?

5. Plans call for a combination of three Type 88 units. What should the rough stud opening be if the individual sash opening is 3'-0" x 3'-2"?

B. Refer to Woodco casements, pages 78, 79, and 80.

1. What is the rough stud opening needed for a single #210 casement?

2. What masonry opening is required for a #412 window?

3. Four #310 windows are to be installed as a multiple unit. What rough stud opening does this unit require?

4. Specifications call for a Woodco #5412C window unit. Does this unit contain any fixed windows? Describe.

5. What opening does the above unit require if the construction is brick veneer?

C. Refer to Andersen Flexivent® units, page 81.

1. A triple #214, arranged horizontally, is to be installed in a frame house. What rough stud opening is needed if subsill is used?

2. If a Flexiview® 3849 and a Flexivent® 3818 are installed as a picture window with a hopper beneath it, what masonry opening is required?

3. Nine #3814 Flexivents® are to be used in a combination three units high and three units wide. What rough stud opening is needed?

4. What is the difference between awning and hopper-type windows?

D. Refer to a Curtis catalog. If none is available, your instructor may substitute appropriate types of windows from available catalogs.

1. What is the rough stud opening required for a C-2772 window that is 2'-0" x 3'-2"?

2. What masonry opening is required for a C-2752 window with glass size 20" x 20" with F-77 mold?

3. A triple C-2671 awning window unit, Type 301, is specified for a brick veneer house. Using a band mold other than F-77, what masonry opening is required?

E. Answer the following questions, referring to figure 12-9, page 82.

1. Do any of the details show a meeting rail? If so, identify.

2. Does the end of the stool project beyond the inside casing?

3. Which views were used to answer the preceding question?

4. Is a box frame required for a picture window? Explain.

5. What are the side members of a sash called?

6. Is the stone sill the same length as the width of the window?

7. What provision on the door jambs of a door frame is taken to prevent the door swinging through the frame?

8. What types of door frames are described?

9. Name three descriptive terms which are common to both doors and sash.
   (1) _____ (2) _____ (3) _____

10. Name three methods by which doors are hung in door frames.
    (1) _____ (2) _____ (3) _____

11. Name two methods of describing lights of glass that are common to sash and doors.
    (1) _____ (2) _____

12. Name three wood members whose terms are common to both sash and doors.
    (1) _____ (2) _____ (3) _____

UNIT 13   DETAILS-STAIRS

OBJECTIVES

- Describe methods of building staircases.
- List those sections of a staircase that are detailed.
- Show construction details of various types of staircases.

Staircases are included on floor plans to show their locations in terms of the surround-

ing partitions. The stair drawings are drawn to the same scale as the floor plan. Therefore, the

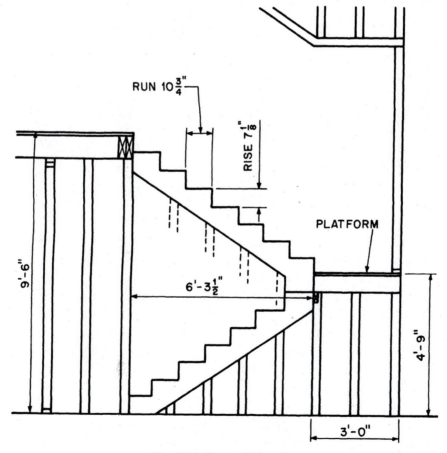

RUN $10\frac{3}{4}"$

RISE $7\frac{1}{8}"$

PLATFORM

$9'-6"$

$6'-3\frac{1}{2}"$

$4'-9"$

$3'-0"$

Fig. 13-1 Stairwell framing

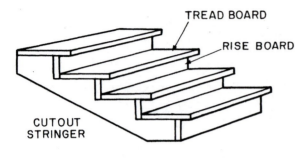

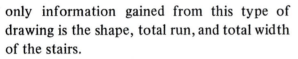

Fig. 13-2 Cutout stringer

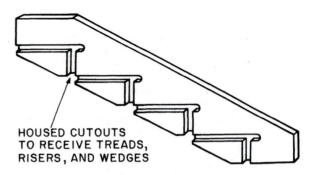

HOUSED CUTOUTS
TO RECEIVE TREADS,
RISERS, AND WEDGES

Fig. 13-3 Housed stringer

only information gained from this type of drawing is the shape, total run, and total width of the stairs.

To determine the exact location and type of platform, supporting partitions, and headroom, the elevation of the staircase is usually shown in detail, figure 13-1.

There are many methods of building staircases. Some staircases have 2″ x 12″ members (called *stringers*) cut to the contour of the tread and riser to support the treads and risers, figure 13-2. Another type of stair is composed of finish stringers into which the ends of the treads and risers are housed, glued, and wedged, figure 13-3.

Some types of staircases are fully enclosed on both sides by partitions, some are partially enclosed, and others are not enclosed but are exposed to handrails. To show these various types of staircases clearly, their details are drawn to a larger scale. This helps to determine the intent of the drawings and to avoid misunderstanding.

The scale of the floor plan is also too small to show the exact overall width of the staircase. It would be impossible to show whether the overall width of the staircase is to be taken as the distance between the unfinished faces of the well hole partitions or between the finished faces of the partitions. To solve this problem, dimensions are given between the

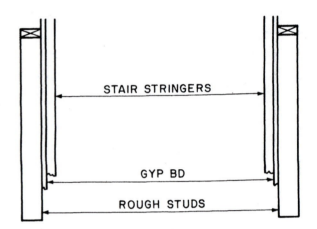

Fig. 13-4 Dimensioning stair details

stair stringer, between the interior wall finish, or between the studs, figure 13-4.

In many instances, the staircase leads from the first floor to the floor above it, such as in the case in a two-story or split-level house. This type of staircase is generally exposed to a handrail on one or both sides. This requires detail drawings of the rough framing of the stair carriage and the detail of the finish stairs and handrails, figure 13-5.

Problems caused by the many variations of stair types and complications that arise in stair measurement are best solved by a thorough knowledge of making and reading detailed stair drawings.

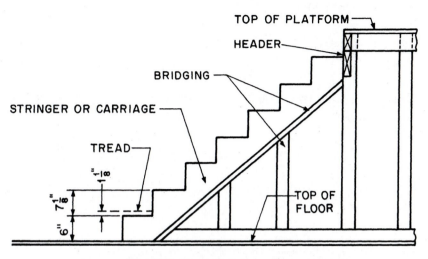

Fig. 13-5 Framed stair carriage

ASSIGNMENT

A. Answer the following questions, referring to the drawings of the house, Sheets 1/7 through 7/7, in the back of the book for questions 7–15.

1. What type of drawing(s) is required to show the headroom of the staircase on a floor plan?

2. What type of drawing(s) is required to show the overall rise and run of the staircase?

3. What type of drawing(s) is required to show that a staircase is enclosed on both sides from floor to ceiling?

4. What type of drawing(s) is required to show the method of supporting the treads and risers to the stringers?

5. Which of the three methods of building staircases mentioned in this unit is the best? Explain your answer.

6. Name, in order of importance, five of the most important measurements required to lay out a finish staircase.

7. What is the total rise of the stairs leading from the basement to the first floor?

8. What is the individual step rise?

9. What is the total run?

10. What is the individual tread run?

11. The stairwell width is 3'-4''. What is the length of the tread and riser boards if the stringer is 1 1/16 inches thick and housed 3/8 inch deep?

12. What is the total rise of the steep staircase leading from the platform to the attic floor?

13. Name the most important reason why this attic stair needs to be steep.

14. Approximately how many risers and treads are put in an attic stair of this type?

15. What is the individual rise and run?

B. Make a scaled drawing showing the rough frame carriage of a staircase with one side exposed from floor to ceiling. Draw the exposed side elevation, using the detail drawing of the stairs on Sheet 3/7 in the back of the book as a working model. Show all important dimensions. Scale 1 1/2'' = 1'-0''.

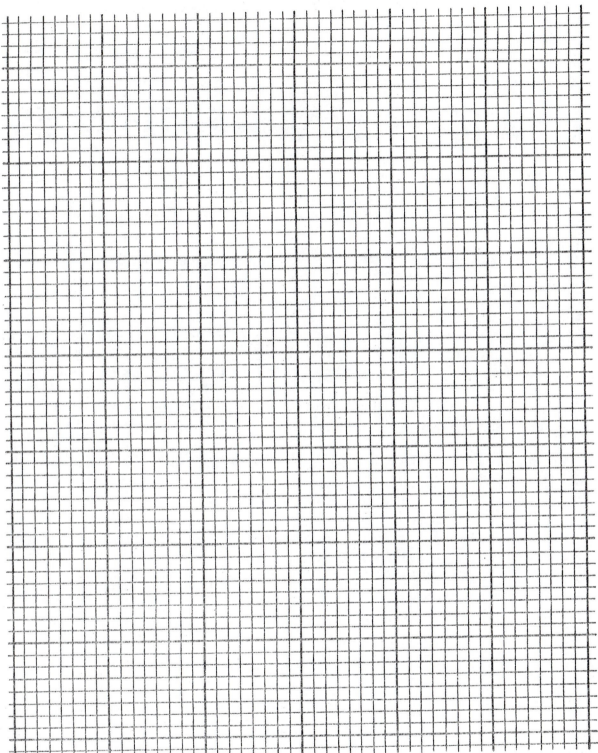

UNIT 14 DETAILS – FIREPLACES

OBJECTIVES

- Explain methods of providing fireplace mantels and bookcases as required by drawings.
- Discuss procedures used to obtain the information from drawings that is required by the mill.

In many cases, fireplaces require no mantel shelf or millwork. The face of the fireplace is simply finished in brick or stone to compliment the decor of the interior.

In other instances, however, more elaborate mantel shelves and bookcase cabinets are required. These might be difficult to build on the job. For example, the fireplace in figure 14-1 might require millwork on the wood trim. The architect then supplies shop drawings so that such trim might be made at the mill. These are made according to the plans and specifications or according to the standardized units as manufactured by various mills.

In the detail drawing of the first-floor fireplace, figure 14-2, the plan view shows that the face of the brickwork of the fireplace is flush with the face of the plastered wall. This continues only to the bottom of the mantel shelf. At this point, the brickwork is recessed to provide space for the installation of 2 x 4 studs, plasterboard, and plaster which finishes flush with the plastered wall above the mantel shelf.

The recessed section of the chimney, being the full width of the chimney, is clearly shown in Section A-A, figure 14-2.

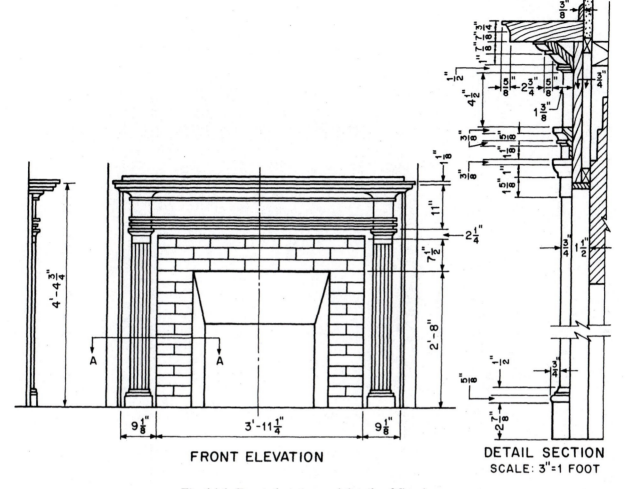

FRONT ELEVATION

DETAIL SECTION
SCALE: 3"=1 FOOT

Fig. 14-1 Front elevation and details of fireplace

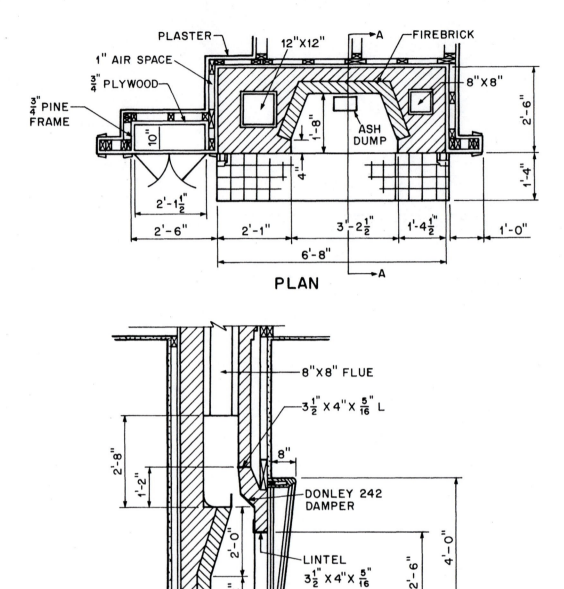

PLAN

SECTION A-A

Fig. 14-2 Detail of first-floor fireplace

ASSIGNMENT

A. Answer the following questions, referring to the drawings of the house, Sheets 1/7 through 7/7, in the back of the book.

1. How many doors are required for a bookcase?

2. What is the approximate size of the doors?

3. Is there a wood door jamb on which the doors are to be hung?

4. Are there any doors enclosing the upper shelves?

5. Of what material is the mantel shelf made?

6. Is the tile floor flush with the living room floor?

7. At what point of the elevation view is the section taken?

8. What is the meaning of the closely spaced hatch marks surrounding the fireplace opening in the plan view?

9. What is the smallest flue shown in the plan view?

10. How many flues are shown in the elevation view of the chimney in the recreation room?

11. What items do the broken lines in the elevation view of the chimney in the recreation room indicate?

12. Is the face of the furring flush with the face of the chimney?

13. What size are the furring strips?

14. What size is the V-joint pine paneling?

15. Is the face of the baseboard on the same line as the face of the pine paneling?

16. Is the 12″ x 12″ flue lining in the recreation room chimney surrounded with the same thickness of brick walls as the same lining in the living room chimney?

17. Is the flue lining for the living room fireplace shown in any other place than in section?

18. What does the broken line circle in the elevation view of the chimney in the recreation room indicate?

19. Is there any wood trim shown around the fire opening in the fireplace in the recreation room?

20. Is there a mantel indicated for the recreation room fireplace?

B.   Sketch section A-A, figure 14-2, in the plan view using a scale of 3″ = 1′-0″.

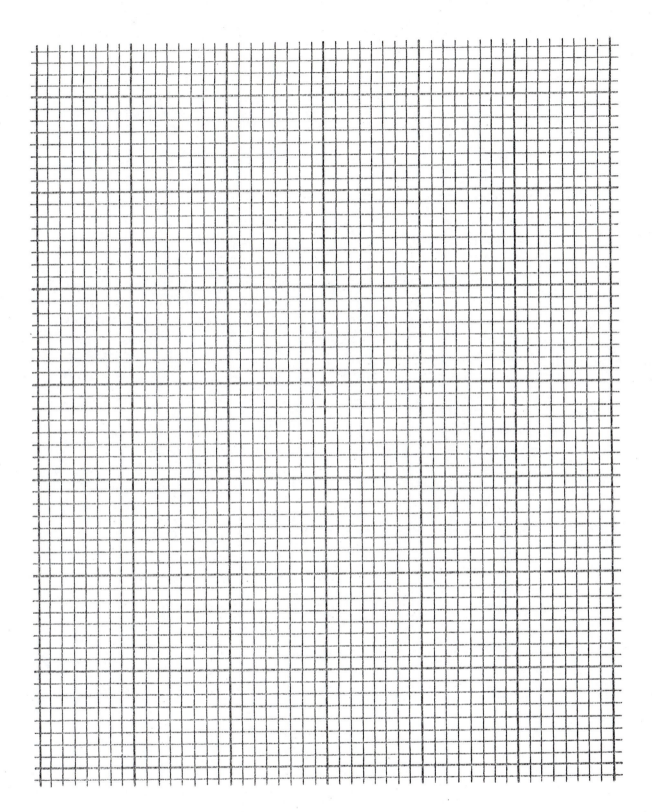

UNIT 15   DETAILS-CABINETS

OBJECTIVES

- List items included in detail framing of cabinets.
- Describe a typical detail drawing for custom-made cabinets.
- Explain the principles of unit cabinet construction.

The floor plan of the kitchen in Print No. 3/7 shows the arrangement of the various types of cabinets on the kitchen floor adjacent to the walls. The location of the fixtures, such as the sink, range, refrigerator, electric fan, and clock, are all important in respect to the general convenience of the kitchen. They must be definitely located in order to give the location of service lines required by these fixtures.

The function of this type of drawing is to show in general the area taken by these cabinets in respect to the total area and shape

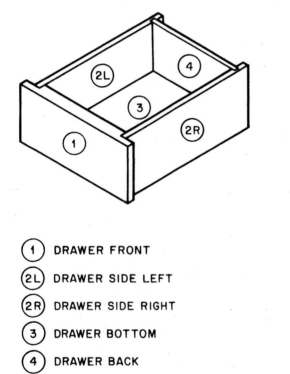

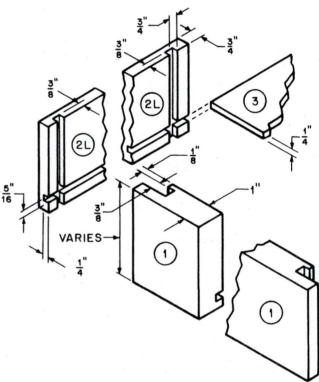

(1) DRAWER FRONT

(2L) DRAWER SIDE LEFT

(2R) DRAWER SIDE RIGHT

(3) DRAWER BOTTOM

(4) DRAWER BACK

Fig. 15-1 Cabinet details

of the kitchen floor. The location of doors and windows and the available area for kitchen workspace and movable furniture, such as table and chairs, is also indicated.

It may be seen that this type of drawing is similar to the floor plan of the staircase. It shows no detailed information regarding the height of the counter shelves, vertical location of the wall cabinets, and size, type, or swing of the cabinet doors. Here, again, elevation drawings are required to show the vertical size, location, and position of floor and wall cabinets, doors, drawers, fixtures, and outlets.

Print No. 7/7 shows a general elevation of these cabinets in their locations on the various walls of the kitchen. However, neither the floor plans nor the elevations provide informa-

tion regarding the definite type and construction of such items as the wall cabinets, doors, counter shelves, drawers, and drawer slides.

Elevation drawings, in most cases, are sufficient where prefabricated stock cabinets in steel or wood are to be used. If cabinets are custom-made, typical illustrations of the required cabinet drawings showing more definite detail would be similar to Section D-D, Print No. 4/7.

This method of detailing, when applied to cabinet detailing, may be further expanded by circling any required section of the section drawing and making this drawing full size. Figure 15-1 illustrates how this method is used to show the construction of cabinet drawer joints.

ASSIGNMENTS

A. Answer the following questions, referring to the specifications in the Appendix and the house plans, Sheets 1/7 through 7/7, in the back of the book.

1. How far down is the ceiling dropped for the wall cabinets?

2. How deep are the base cabinets at floor level?

3. Is the range built in or is it a separate, movable unit?

4. How many drawers are there in the kitchen cabinets?

5. What is the distance from the top of the counter shelf to the top of the floor at the sink?

6. What is the distance from the top of the floor to the top of the counter-top at the range?

7. What is the height of the kick space?

8. What is the size of the stiles?

9. What is used as a backsplash?

10. What type of catches are used on the cabinet doors?

11. What material is used to construct the drawer bottoms?

12. What material is used to construct the drawer sides?

13. What material is used to construct the kitchen cabinets?

B. Sketch the back of the drawer in figure 15-1. Include all necessary dimensions. Use style and scale similar to figure 15-1.

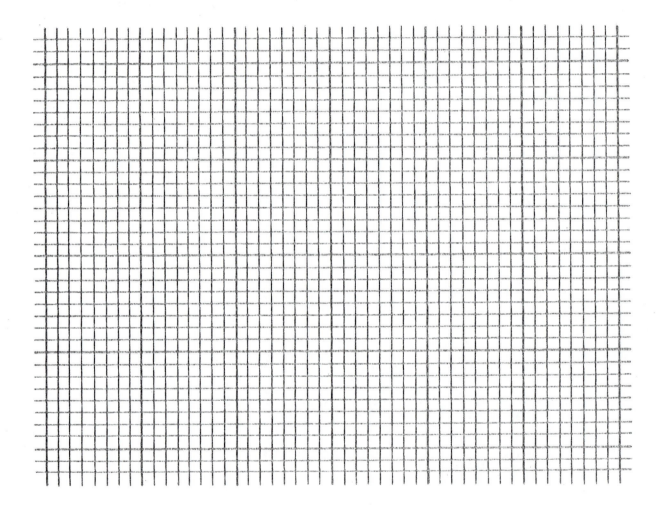

C. Make an isometric sketch of a wall cabinet on the northeast wall (Sheet 7/7) in the back of the book. Show shelves and method used to support them. Scale 1 1/2″ = 1′-0″.

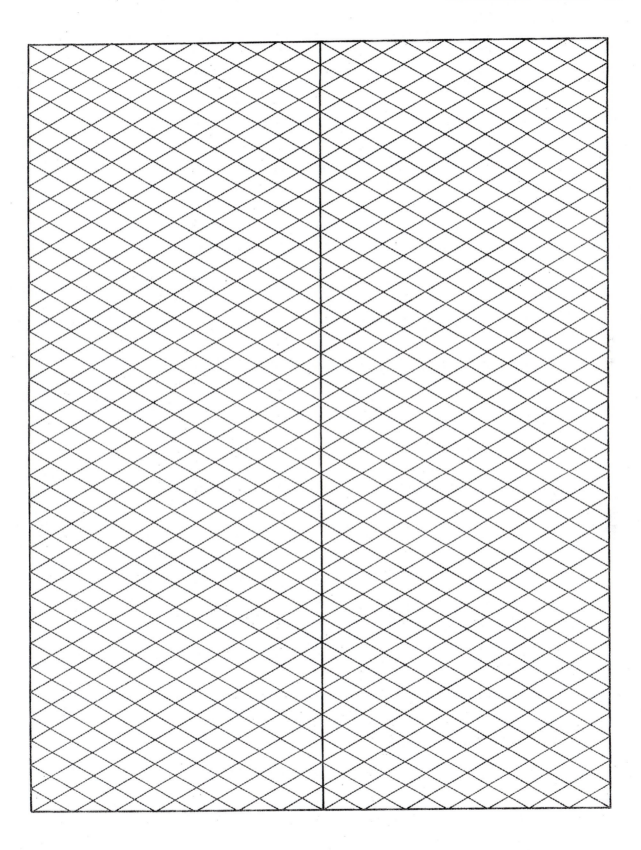

D.   Make a sketch of the cabinet on the northeast wall (Sheet 7/7) in the back of the book. Show the method of framing the doors into the door openings and against the stiles and rails of the lower cabinet. Scale: full size.

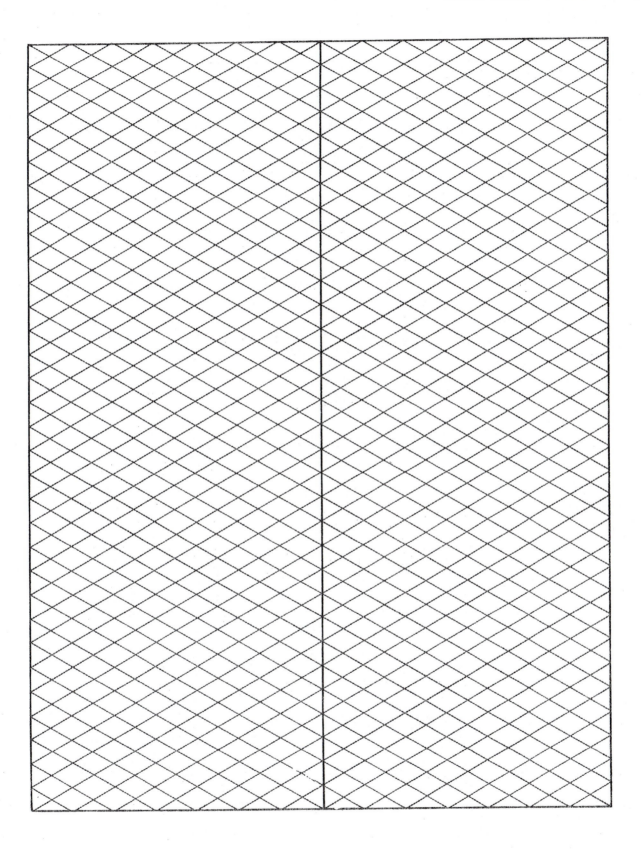

UNIT 16   THE HEATING AND AIR-CONDITIONING PLAN

OBJECTIVES

- Describe two ways in which buildings are heated.
- Describe one way in which buildings are cooled.

In its simplest form, the heating and air-conditioning plan is a floor plan with climate control equipment, figure 16-1. This equipment includes ducts, furnace, piping, plenum, and registers. (See figure 1-3 for air duct symbols.)

There are several different methods of heating and cooling buildings. Some of the more popular methods are electric heating, forced warm air, hydronic systems, and central systems.

Electric Heating

An electric heating system is an efficient means of heating a home. Some of the more popular electric heating systems are baseboard units, cable systems, central furnaces, and heat pumps.

Baseboard heating units are designed to fit next to the base of a wall that might experience a large heat loss. These units can be individually controlled, or they can be controlled by a central thermostat.

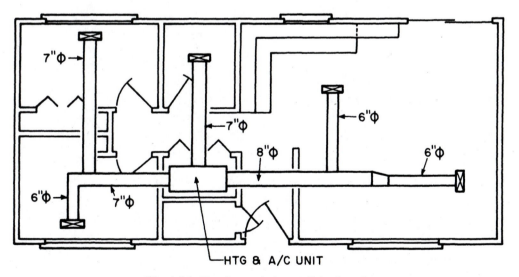

Fig. 16-1 Heating and air-conditioning plan

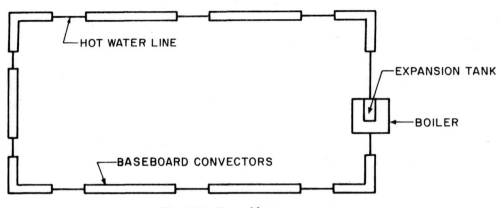

Fig. 16-2 Forced hot-water system

An *electric cable system* is created by embedding electrical wires in the floor or ceiling. The system operates by reflecting the various heat rays from the surfaces of the room.

A *central electric furnace* system operates by heating air in a furnace. It then is distributed throughout the house in a system of ducts.

A *heat pump* operates by reversing the conventional refrigeration cycle. The evaporation coils in the unit cool the air outside, and the condenser coils warm the air inside. A network of ducts is used to distribute the air throughout the house.

Hydronic Systems

A hydronic system is nothing more than a forced hot-water system, figure 16-2. The water is heated in a large boiler and then pumped through a series of pipes and into convectors. After the hot water passes through the convectors, it is returned to the boiler and reheated. The process is repeated time and time again.

Forced Warm Air

A forced warm-air system operates by heating air in a furnace and distributing it throughout the house in a system of ducts. To release the warm air, the ducts are fitted with registers. Ducts can be placed in almost any location, but the system is most efficient if the registers are placed near the floor and windows.

Air-Conditioning Systems

Air-conditioning systems are usually classified as being either a central or unit system. A central system uses the same distribution system as the forced warm-air system. Unit systems are usually placed either in a window or in an opening in a wall.

ASSIGNMENT

A.   Make a sketch of two bedrooms back to back. To heat the two rooms, sketch in a hydronic system.

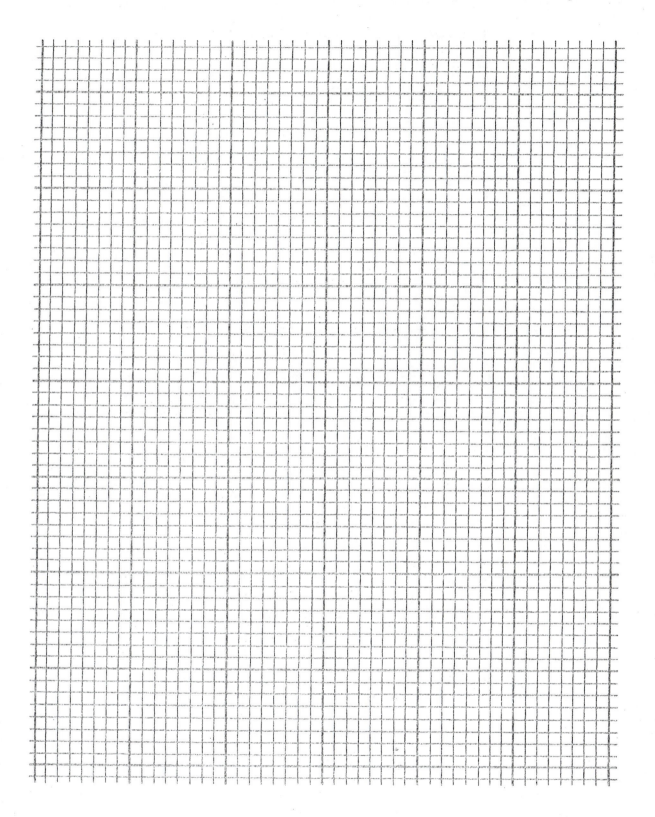

B.   Sketch a forced warm-air system in the two-bedroom home in figure 16-3.

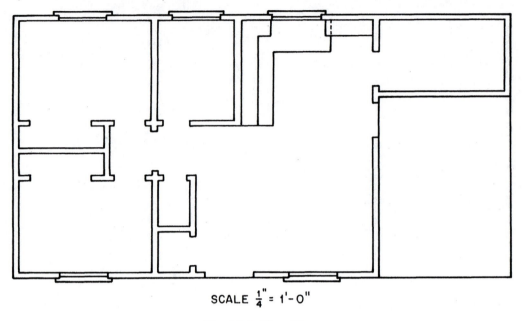

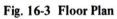

SCALE $\frac{1"}{4}$ = 1'-0"

Fig. 16-3 Floor Plan

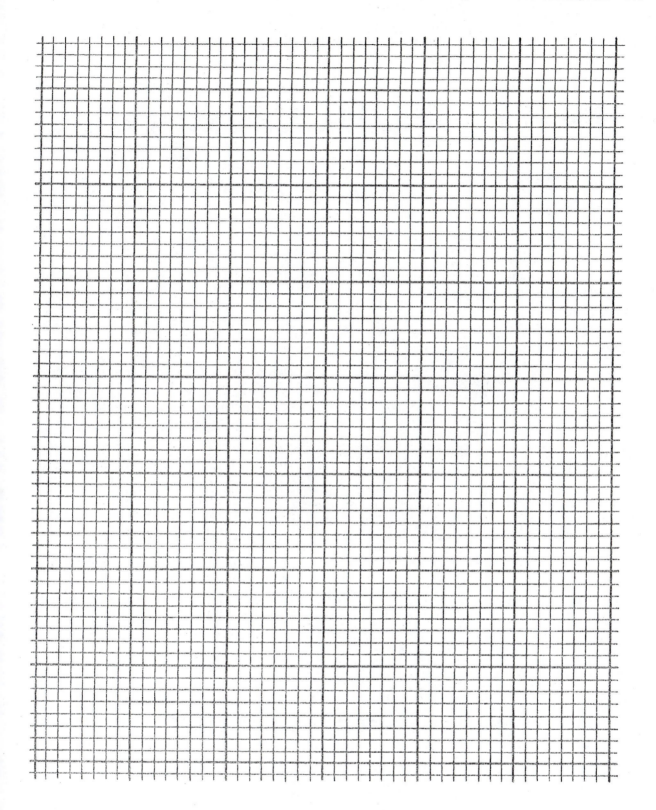

UNIT 17 THE PLUMBING PLAN

OBJECTIVE

- Describe features of a typical plumbing plan.

There are now local, state, and federal laws which set strict minimum health standards for plumbing installation. To meet these health standards, a set of blueprints are drawn to scale with the location of each fixture properly positioned, figure 17-1. The routing and sizing of the various hot and cold water lines are also located on the working drawing. Symbols on the drawing identify the systems and show the location of the valves and fittings. (See figure 1-2 for a list of plumbing symbols.)

The Building Drain and Building Sewer

The *building drain* and *building sewer* are the lowest horizontal pipes in a drainage system, figure 17-2. They receive and discharge waste from soil and waste stacks. The building drain is located under the building to a point 3 feet from the outside edge of the building. The building sewer starts 3 feet from the outside edge of the building and extends to the sewer.

Soil and Waste Pipe

A *soil pipe* is a vertical portion of the piping system, figure 17-3. It carries the discharge of water closets, urinals, or fixtures having

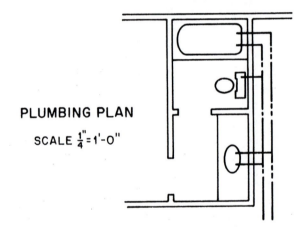

PLUMBING PLAN

SCALE $\frac{1''}{4}$=1'-0"

Fig. 17-1 Plumbing plan (bath)

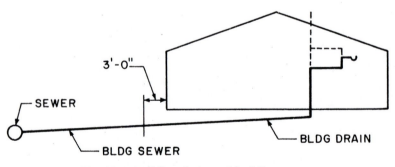

Fig. 17-2 Building drain and building sewer

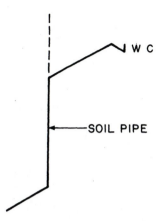

Fig. 17-3 Soil pipe

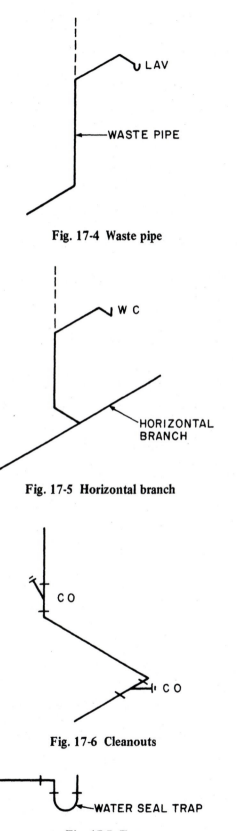

Fig. 17-4 Waste pipe

similar functions. A *waste pipe* is similar to a soil pipe, with the exception that the waste pipe does not carry waste from water closets or urinals, figure 17-4.

Soil Branch

A *soil branch* is a horizontal branch of the piping system that receives the discharge of water closets, figure 17-5.

For easy and accessible cleaning, each soil branch should have an adequate number of cleanouts. Cleanouts are usually placed at the end of the branch furthest from the soil stack and where the soil branch changes direction, figure 17-6.

Traps

A *trap* is a bend in a soil or waste line that is filled with water, figure 17-7. Traps are used in plumbing systems to prevent gases from backing up and spilling into the residence.

Vents

Vents prevent traps from siphoning. Some of the more typical vents are the common vent, stack vent, loop vent, circuit vent, branch vent, and continuous vent, figure 17-8.

- A *common vent* serves two fixtures that are placed back to back.
- A *stack vent* is a continuation of a soil or waste stack.

Fig. 17-5 Horizontal branch

Fig. 17-6 Cleanouts

Fig. 17-7 Trap

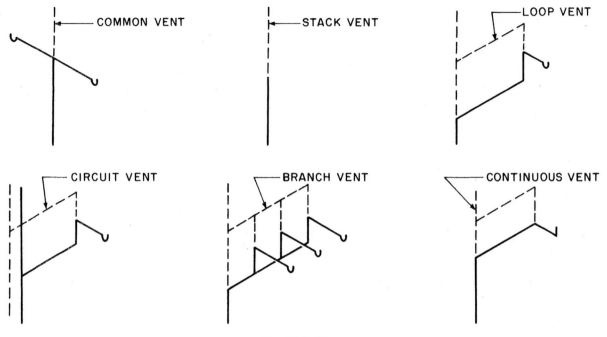

Fig. 17-8 Vents

- A *loop vent* serves two or more fixtures and extends from in front of the last fixture to the stack vent.

- A *circuit vent* serves two or more fixtures and extends from in front of the last fixture to the vent stack.

- A *branch vent* connects one or more individual vents with a stack vent or vent stack.

- A *continuous vent* is a continuation of a soil or waste stack.

Water Supply Systems

A typical water supply system is constructed of fixtures, valves, and a series of lines. There are several different types of plumbing fixtures, but the most common are lavatories, tubs, and water closets. Lavatory bowls are available in various sizes and shapes. They are usually made of vitreous china, stainless steel, or fiberglass.

A bathtub usually measures 2'-6" x 5'-0", although there are other sizes available. They are usually constructed in one of three ways: molded cast iron with a porcelain enamel surface, formed steel with a porcelain enamel surface, or fiberglass.

Water closets are either floor mounted or wall mounted. Most water closets are made of vitreous china.

A hot-water system is an important part of any water supply system. It includes the hot-water tank and a series of lines to feed the various fixtures.

ASSIGNMENT

Answer the following questions.

1. What is the name of the lowest horizontal pipe in a drainage sytem?

2. Where does a building sewer start?

3. What is a soil pipe?

4. How does a waste pipe differ from a soil pipe?

5. Where are cleanouts located?

6. Why is a trap used?

7. Describe the function of a common vent.

8. How does a loop vent differ from a circuit vent?

9. Why are vents used?

10. List six types of plumbing vents.

UNIT 18   THE ELECTRICAL PLAN

OBJECTIVES

- List six items shown in a typical electrical plan.
- Describe the location of electrical outlets and fixtures in buildings.

An *electrical plan* is a plan view drawing similar to the floor plan, figure 18-1. It shows the location of switches, lighting fixtures, circuits, special electrical equipment, the meter, and the distribution panel. (See figure 1-4 for electrical symbols.)

Electrical Outlets

Electrical outlets are indicated on the plan by a small circle with short lines perpendicular to the stud wall. In living areas, these outlets usually are placed 6 to 8 feet apart. Other areas, such as the hall, foyer, carport, and storage room, usually are provided with at least one electrical outlet.

For outdoor use, a weatherproof outlet usually is placed on patios, porches, garages and carports. These outlets are designated by placing the letters *WP* by the outlet.

Electrical Fixtures

Each area of a house should be lighted by at least one electrical fixture. The size and shape of a room usually dictates the number and location of fixtures. Bedrooms usually have one fixture located in the center of the ceiling. Kitchens usually have two lights, one in the center of the room and one directly over the sink. Halls should have lighting fixtures 15 feet apart to be controlled at each end of the hall. Each closet should have a light fixture operated by a switch or pull cord.

Indicators

The symbol for an incandescent ceiling fixture is a circle with four short lines protruding from it. The symbol for a wall switch is

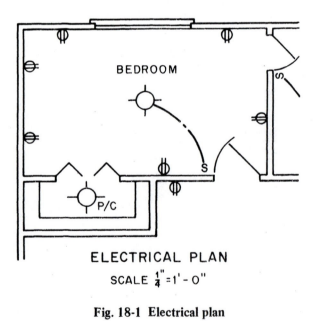

ELECTRICAL PLAN

SCALE $\frac{1}{4}$" = 1' - 0"

Fig. 18-1 Electrical plan

the letter *S*. To connect the ceiling fixture and switch, a curved centerline is used. The switch is located on the door knob side of the door.

If the fixture is such that it will be turned off or on in more than one location, a three-way switch is used.

ASSIGNMENT

Sketch the electrical fixtures and outlets in the floor plan in figure 18-2.

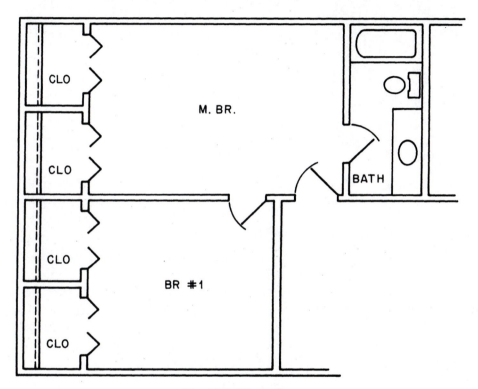

Fig. 18-2 Floor plan

UNIT 19 QUANTITY AND MATERIAL LISTS

OBJECTIVES

- List the reasons for making quantity and material lists.
- Explain the organization of quantity and material lists.
- List some methods of taking off quantities and materials from building drawings.

In most instances, quantity and material lists are used to determine the approximate cost of the project. These lists are made by studying the plans and specifications and then selecting the proper amounts, descriptions, costs, and means of obtaining materials. In many instances, the lack of a list of materials has been the reason for delay in ordering materials. This may cause delay in the construction work, may increase the cost of materials on a rising market, and may delay delivery to the job site because the manufacturer or mill required a certain lapse of time between order and delivery.

Quantity and material lists provide an excellent means of applying a first-hand knowledge of the entire construction, assembly, and requirements of the completed project. They require thorough and correct interpretation of the drawings. The organization or divisions of the list should follow in sequence according to the procedure of building the project, or the sequence of materials listed in the specifications.

Many estimators have sheets prepared in columns. Each column is labeled by item number, quantity, description of material, cost, and other notations. These headings are recorded in a logical order. From these lists, the totals of the materials that are grouped together are compiled and priced in quantities as sold by the builders' supply houses.

Itemized quantity lists are also valuable to have in case a change involving an addition or deduction of materials or costs is made in the construction. In such cases, the list acts as a convenient guide.

The mathematics involved in the measurement of surfaces, volumes, materials, and practical items of construction cannot be divorced from the preparation of quantity and materials lists. A carpentry-related mathematics book, such as *Practical Problems in Mathematics for Carpenters,* is suggested as a reference for the work of this unit.

EXCAVATION

Example: Calculate the volume of earth to be excavated for the footings in the enclosed set of plans.

1. Determine from the drawings and specifications the cubic volume to be excavated.

2. Apply the principles of cubic measure in *Practical Problems in Mathematics for Carpenters.*

CONCRETE BLOCK AND BRICK UNITS

Concrete blocks are building units with common sizes of 8 inches, 10 inches, and 12 inches in thickness; 8 inches in height; and 16 inches in length. In determining the square feet of a wall surface, only the height and length of the block are involved.

Example: Find the number of concrete blocks in 100 square feet of wall surface.

1. 8" (height) x 16" (length) = 128 sq. in. in one block
2. 100 sq. ft. x 144 sq. in./sq. ft. = 14,400 sq. in. in wall surface
3. 14,400 sq. in. ÷ 128 sq. in. = 113 units in 100 sq. ft. of wall surface (approximately)

This holds true if the wall is 8, 10, or 12 inches thick and the units are 8, 10, or 12 inches thick.

The same procedure is followed in figuring brick or other masonry units. If a brick wall is 4 inches thick (the width of one brick), there are approximately 7 1/2 bricks in one square foot of wall surface. If the wall is 8 inches thick, multiply 7 1/2 x 2 (bricks thick) = 15 bricks. A wall 24 inches thick requires 7 1/2 x 6 or 45 bricks per square foot of wall surface.

> *Note:* When taking measurements for quantities, (1) always take exact outside measurements, (2) do not count corners twice, and (3) make deductions in full for all openings.

MORTAR

Mortar quantity is figured according to the thickness of the joint. Divide the number of brick units required in the wall by 1000 and multiply by the following factors.

PER CUBIC FOOT OF MORTAR
(including waste)

Thickness of Mortar Joint	Multiplying Factor
1/4 inch	10
3/8 inch	15
1/2 inch	18
5/8 inch	22

To find cubic yards of mortar, divide this quantity by 27. Allow about 3.25 cubic feet of mortar for every 100 square feet of concrete block wall surface with 3/8-inch mortar joint.

CONCRETE MIX

Due to the different proportions of cement, sand, and coarse aggregates in concrete, individual materials are figured as follows. Multiply the cubic yards of concrete required (in the mix desired) by the factors for cement in number of sacks, sand in cubic feet, and coarse aggregate of 3/4-inch diameter in cubic feet. One sack of cement = 1 cubic foot; four sacks of cement = 1 barrel (bbl.)

Example: Assume that 10 cubic yards of concrete in 1:3:4 mix are required. For number of sacks of cement, multiply 10 x 1.26 = 12.6 bbls; for sand, multiply 10 x 0.58 = 5.8 cubic yards; for aggregate, multiply 10 x 0.77 = 7.7 cubic yards.

PER CUBIC YARD OF CONCRETE
(including waste)

Mix	Factor for Cement in Bbls.	Factor for Sand in Cu. Yds.	Factor for Aggregate in Cu. Yds.
1:2:3	1.70	0.52	0.77
1:2:4	1.46	0.44	0.89
1:2:4 1/2	1.36	0.42	0.93
1:3:4	1.26	0.58	0.77
1:2 1/2:5	1.19	0.46	0.91

CEMENT AND PLASTER COATS

Concrete block walls, especially below grade, are sometimes plastered on the outside with a dampproofing cement plaster of 1/4, 1/2, 3/4, or 1-inch thickness and a mix of 1:1, 1:1 1/2, or 1:2 parts of cement and sand.

Example: Estimate the quantity of cement, plaster, or stucco needed for a job.

1. Find the square feet of wall surface to be covered.

2. Divide this figure by square feet of wall surface the required thickness of plaster can cover (see table below).

3. The resulting figure is the cubic yards of plaster required.

Use the table below to find the approximate number of square feet of wall surface that 1 cubic yard of the various thicknesses of plaster coat can cover.

PER CUBIC YARD OF PLASTER
(1:2 Mix)

Thickness of Coats in Inches	Square Feet of Wall Surface Covered
1/4 inch	1,296
1/2 inch	648
3/4 inch	432
1 inch	324

NOTE: Do not deduct for openings unless they are more than 16 square feet in area.

Example: Apply a 1/2-inch coat of mortar on an 8' x 10' wall.

1. 8' x 10' = 80 sq. ft. to be covered.

2. 80 ÷ 648 = 0.123 cu. yd. of mortar needed.

Use the following table to find the amount of cement and sand required for 1 cubic yard of mortar of the various proportions.

PER CUBIC YARD OF MORTAR

Mix Cement Sand	Sacks of Cement	Sand (Cu. Yds.)
1:1	19.5	0.72
1:1 1/2	15.5	0.86
1:2	12.5	0.95

Example: Find the amount of mortar required by 0.123 cu. yd. (in a 1:1 ratio).

1. 0.123 x 19.5 = 2.388 or 2.4 sacks of cement.

2. 0.123 x 0.72 = 0.0885 or 0.09 cu. yd. of sand (0.09 cu. yd. x 27 cu. ft./cu. yd. = 2.43 cu. ft.)

FRAMING MATERIALS

Framing members come in stock sizes and lengths. They are itemized in number and lengths required and totaled in number of board feet. A Table Reckoner, which shows the number of board feet in standard size pieces of lumber, may be used.

SILLS, GIRDERS, AND PARTITION PLATES

Find the lineal feet required for a single member and multiply the length by the number of members needed to make up a sill or girder.

Example: How many pieces and board feet are required to make a two-member sill, 2" x 6", 200 lineal feet long?

1. 200 x 2 = 400 lineal feet of 2" x 6"

2. 20 pcs. 2" x 6" x 16'-0" = 320 lineal ft.
   8 pcs. 2" x 6" x 10'-0" = +80 lineal ft.
   400 lineal ft.
   $$\frac{2 \times 6 \times 400}{12}$$

3. 400 lineal feet 2" x 6" = 400 board ft.

FLOOR JOISTS

The required number of floor joists in a given floor area depends upon the spacing between the joists. The length of the joists depends upon the overall distance between the joist supports.

Example: A floor platform 20'-0" long with a 16'-0" span is to be covered with floor joists spaced 12" OC.

1. To find the number of joists required for every foot of the floor platform, divide the spacing (12" in this example) into 12 (in./ft.). Therefore 12 ÷ 12 = 1 joist for every foot of the 20 ft. long platform, which gives a total of 20 joists.

2. Add 1 joist to every span to allow for the end joist. Thus, 20 + 1 = 21 joists 16 ft. long are required.

If the joists are spaced 16" OC, divide 16 into 12 = 3/4. The 20' length is multiplied by 3/4 (20 x 3/4 = 15) + 1 joist for the end = 16 joists required. If the spacing is 20" OC multiply by 3/5 (12 ÷ 20 = 3/5). If 24" OC, multiply by 1/2 (12 ÷ 24 = 1/2), and so on for any width spacing. Remember to add the extra joist.

FLOOR BRIDGING

The number of linear feet of bridging required is found by using the formula:

$$B = L \times N \times 3$$

B is the number of linear feet of cross bridging, L is the length of the building in feet, and N is the number of double rows of bridging. This formula allows for waste.

NOTE: This method does not allow for double joists under bearing partitions, stairwells, and other openings where header and trimmer joists are used.

SUBFLOORING, SHEATHING, AND ROOFING BOARDS

These areas are usually covered with plywood or other sheet material sold in 4' x 8' pieces. To find the number of sheets, find the square-foot area and divide by 32.

STUDDING – DRAFT STOPS

Estimating the required number of studs in a partition or sidewalls depends upon the spacing of studs, the number of corners, and the number of openings.

Example: Assume that a straight partition with studs spaced 16" OC, no openings, and no double corner posts is to be estimated for the number of required studs.

1. The method explained for finding the number of floor joists could be used. If there are considerable openings and corners in the partitions, refer to *Practical Problems in Mathematics for Carpenters.*

GABLE STUDS

When figuring the lineal feet of gable studs required, take half the width of the gable in feet. Multiply this figure by the length of the stud required for the full height of the gable. This gives the lineal feet of studs required if spaced 12 inches OC. For spacing 16 inches OC, use only 3/4ths the figure at right.

Example: If the gable width is 20 ft. and the height of the middle stud is 6 ft., 20/2 x 6 = 60 lineal feet of studding required for spacing 12" OC. For 16" OC spacing, 3/4 x 60 = 45 lineal feet required.

ROOF RAFTERS

In a roof where only common rafters are required, the number of common rafters may be found by the same method as used to find the number of floor joists in a floor platform with no openings. The only difference is that both sides of the roof must be figured in the total rafters required.

Where shed dormer rafters are required, the number required is figured exactly as floor joists.

Where hip or valley jack rafters are required, they are figured by the same method explained for gable studs. Where the roof includes hips, valleys, ridgeboard, and chimney openings, the number of these various rafters are taken off the roof plan.

The lengths of all the above mentioned rafters are found by using the tables on a rafter square.

The remaining items of the average residential building that generally are estimated include the following: wall plates and shoes, rafters, sheathing and roof boards, siding, trim, roof covering, doors and windows, stairs and interior door jambs, finish flooring and paper, and hardware and supplies. For information on estimating these items, refer to *Practical Problems in Mathematics for Carpenters,* or another carpentry-related mathematics textbook.

ASSIGNMENT

Answer the following questions, referring to the floor framing plan prepared for the Assignment in Unit 10, living room and dining room areas.

1. How many regular joists are required?

2. Name the size and length of the joists.

3. How many headers are required?

4. State the length of each header.

From the window schedule, answer the following questions.

5. How many double-hung windows are required?

6. How many multiglazed windows are required?

7. How many casement windows are required?

8. Estimate the board feet of joists required for the wooden deck. Show, in detail, how these figures are derived.

9. Estimate the lineal feet of gable studs required for a building 24 feet wide with a roof pitch 8 on 12. Show all work in arriving at this answer.

10. Estimate the number of concrete blocks required for the residence. After checking these figures, estimate the amount of mortar required for these blocks.

COMPREHENSIVE REVIEW

PART I

The following questions are based on the specifications in the Appendix and the drawings of the house, Sheets 1/7 through 7/7, in the back of the book.

1. Convert the following engineer's dimensions to equivalent feet and inches:
   a. 140.212 feet

   b. 94.114 feet

   c. 200.423 feet

2. If the bench mark elevation is 200.32 feet and the bottom of the foundation footing is 188.86 feet, what is the difference in feet and inches?

3. What is the approximate pitch in feet of the pipe from the house to the septic tank?

4. What is the approximate pitch of the pipe from the septic tank to the field?

5. How is water to be supplied to this house?

6. Suggest a location for the dry well. Give this location in distance and direction from the building.

7. If the drainage field is 2 feet below grade, how many feet is it below the sidewalk?

8. What are the dimensions of the footings under the 12-inch walls?

9. What concrete mix is used for the footings?

10. List the various thickness of concrete walls shown on the basement plan.

11. Give two thicknesses of the retaining walls.

12. Are footings shown under the basement partitions?

13. What is the dimension from the face of the outside garage wall to the center of the #12 window?

14. What are the inside dimensions of the storage room?

15. What is the size of the lintel over the cellar windows?

16. How far does the fireplace project into the furnace room?

17. Eliminating openings, what is the total length of all the basement cinder wall shown?

18. What size gravel is used for fill?

19. Give the distance from the outside face of the front foundation wall to the face of the fireplace.

20. What is the depth of the fireplace opening in the basement?

21. What is the dimension from the front face of the garage wall to the outside face of the front wall of the storage room?

22. What are the overall dimensions of the concrete areaways?

23. What is the purpose of the flue shown in the basement fireplace?

24. What is the dimension from the center of the garage front pier to the face of the right-hand wall of the garage?

25. Give the distance between the centers of the footings within the garage.

26. Are there footings under the basement concrete block partitions?

27. What is the dimension from the outside face of the pump room wall to the outside face of the side garage wall?

28. If the I-beam over the recreation room rests 6 inches on each wall, how long should it be?

29. If the angle iron over window #12 rests 4 inches on each side of the opening, how long should it be?

30. Why is the garage window not shown in the basement plan?

31. What is the distance from the center of window #3 to the outside face of the framed wall of the rear bedroom?

32. What two methods are used to support the terrace slab?

33. List the different wall thicknesses of the pump room.

34. If the concrete blocks in the basement partitions are 8 inches high including joint, how many full courses are there from the top of the floor to the underside of the joists?

35. List the plumbing and heating items shown in the furnace room.

36. How far is the bottom of the garage footing below the garage floor?

37. How thick are the wood plates used on top of the concrete block walls?

38. How are the plates on top of the cinder block walls held in place?

39. At the front wall, section A-A, what is the dimension from the top of the footing to the top of the concrete wall?

40. Where is dampproofing used?

41. What ceilings in the basement are furred?

42. What is the distance from the center of window #5 to the outside face of the framed wall of the garage?

43. What is the height from the top of the garage floor to the top of the wood platform?

44. What is the width of the hinged closet doors (F)?

45. How many window frames are needed on the first floor?

46. What is the size of the first floor fireplace less hearth?

47. How many lights of glass does the breakfast nook window have?

48. How many lights of glass does each garage door have?

49. What is the total number of flues used?

50. What length joists should be bought for the front bedroom floor?

51. What size timbers should be bought for the front bedroom floor joists?

52. How thick are the basement stair treads?

53. Identify the type(s) of frame wall construction used on this job.

54. What size and type of sheathing is used on this job?

55. What is the dimension from the finish ceiling to the top of the first floor joist?

56. Are the walls of this building insulated?

57. Explain the procedure for determining the location of the #4 window in the garage.

58. What is the length of the wall plate from the right wall of the garage?

59. What is the overall height of the first floor partition using actual lumber sizes?

60. Does the front of the fireplace project beyond the face of the living room partition?

61. Approximately how many lineal feet of closet pole are needed for the bedroom closets?

62. How many metal windows are used on the first floor?

63. How many risers are there from grade level to first floor level at the front door?

64. Give the location of three bearing partitions on the first floor.

65. What is the overall width of the #7 window?

66. What is the distance from the center of the front #12 window to the outside of the pump room wall?

67. What type of finish is used on the bedroom walls?

68. Name the three kinds of exterior finish wall covering required.

69. Is the rear wall of the bedroom in line with the garage rear wall?

70. What type of switch is used to control the ceiling light in the kitchen?

71. How far above the finish floor is the top of the window in the front bedroom?

72. There are three fixed windows on the first floor. Where are they?

73. What is the clapboard exposure?

74. There is a closet that has no shelf or closet pole. Where is it located?

75. Are there any pull chain electric fixtures on the first floor? If so, list.

76. How many risers are there from the garage floor to the kitchen floor?

77. Make a list showing the approximate number of pieces, size, and type of vertical siding required.

78. Make a list showing the approximate board feet of clapboards required to cover the total area as shown on the exterior walls.

79. Give the lineal feet of metal gutter required, the lineal feet of 3-inch leader required, and the number and type of elbow fittings required, as shown on the drawings.

80. Figure the number and fraction of squares of shingles required.

81. Give a complete list showing the number of pieces and sizes of all the oak lintels required.

82. How many square feet of flagstone is required for the front walk?

83. What is the total square feet of rubble stone surface to be covered?

84. What is the total yards of concrete required for the poured concrete footings and walls? Give totals for footings and walls separately.

PART II

Refer to the drawings for the Comprehensive Review in the back of the book, Sheets CR-1 through CR-6.

1. What is the height of the kitchen cabinets?

2. What is the height of the vanity cabinets in the bathrooms?

3. What size windows are used in the master bedroom?

4. How many electrical outlets are used in bedroom #2?

5. What is the size of the cased opening between the kitchen and dining area?

6. Where is a 2 - 2x12 beam used?

7. Where is a spotlight used?

8. What size window is used in bedroom #3?

9. What is the size of the room that houses the HT/AC and HWT?

10. What is the depth of the closet in bedroom #3?

11. What is the width of the foyer?

12. What size ceiling joist is used in the living room?

13. What size door is used at the entrance to bedroom #2?

14. What is the distance between the base kitchen cabinet units and the wall units?

15. What is the thickness of the wall behind the bathtub?

16. What scale is used to draw the elevations?

17. What type of roofing material is used to cover the roof?

18. What is the roof overhang?

19. What is the overall width of the house?

20. How are brick wall ties installed?

21. What is the minimum thickness of a concrete slab?

22. What is the width of most footings?

23. How many rebars are used in Detail A?

24. All concrete should be a minimum of _____ thick.

25. What size fascia is used?

26. What is the thickness of the soffit?

27. How far apart are the weep holes?

28. What size are the anchor bolts?

29. What is used for base flashing?

30. How are interior plates secured?

31. What is used for windbracing?

32. What is the roof pitch?

33. What centers are the studs placed on?

34. What is used for a ridge?

35. What is the distance from finished floor to grade line?

36. What is the minimum distance from the finished floor to the bottom of the footing?

37. What is placed directly under the concrete slab?

38. What are used for lookouts?

39. What size vents are placed on the soffits?

40. What is the size of the front porch?

41. How far should the chimney footing extend into undisturbed earth?

42. Describe the reinforcement used in the chimney footing.

43. What is the width of the carport?

44. What is the depth of the carport?

45. What is used to cover gable ends?

46. What is the finished floor elevation?

47. In what direction does the house face?

48. How many square feet are in the lot?

49. What are the two distances from the property line to the house?

50. How wide is the sidewalk?

51. How are joists connected to a beam?

52. How is brick supported over an opening?

53. What is used to close the opening between the brick and the jamb?

54. What is used as a vapor barrier?

55. What material is used to finish the ceilings?

56. Describe window #A.

57. Describe door #2.

58. How many electric outlets are there in the family room?

59. Where are the cased openings located?

60. How high is the raised hearth?

61. Who is responsible for sizing the ducts?

62. What is the height of the cabinets in the utility room?

63. What type of ceiling is used in the family room?

64. Where is the folding stair located?

65. Describe the columns.

66. Where is a #2 door located?

APPENDIX

SPECIFICATIONS

Note: These specifications refer to the accompanying residential drawings.

SPECIFICATIONS FOR RESIDENTIAL BLUEPRINTS

SECTION 1 – PREPARATION OF SITE

Article 101 – *Clearing Site*

1. The contractor shall clear the site, removing any trees, stumps, and roots which would interfere with the building operations. All such materials and rubbish shall be removed from the premises.

2. During the building operations all precautions must be taken by the contractor to protect all trees that are to remain. Those near the building shall be protected by wooden guards.

Article 102 – *Laying Out Work*

3. The contractor shall lay out the building accurately upon the site following the marks laid out by the surveyor, and shall be responsible for the erection of all profiles or batter boards, setting same true and level.

SECTION 2 – EXCAVATION

Article 201 – *Basement*

1. The contractor shall do all the required excavation as necessary for basement areas, foundation walls, footings, piers, etc., as shown on the drawings. Excavations shall be carried 2 feet beyond walls. Excavations shall be carried to depths as shown on drawings. All top soil shall be removed from work area and deposited where directed. All excess earth, not needed for fill, shall be removed from the premises.

2. If rock is encountered, the removal of same will be considered as extra work under this contract. A set price per cubic foot for excavating rock shall be included in the bid for the excavation and used as a basis of payment for such work, in addition to the regular bid for earth removal.

3. The bottoms of all excavations shall be left level and smooth, to proper depths as shown. Should water be encountered during excavation, the contractor shall be responsible for elimination of same, to prevent the holdup of future form of concrete work.

4. All excavation around foundation walls shall be made large enough to allow the placing of a 4-inch vitrified drainage pipe as shown at footing level. The contractor will lay this pipe with butted joints with a pitch of not less than 1/4-inch to the foot. The trench for this pipe will be filled with 1 inch gravel not less than 12 inches above said pipe. The line of pipe shall be run to some safe distance from the building where the water can be allowed to soak away by means of a dry well.

Article 202 – *Dry Wells*

5. Excavate for and build a dry well for footing drainage. Well shall be 6 feet by 6 feet and 5 feet below cellar bottom and twenty feet away from the building. Refill with stone to within 3 feet of grade; cover with creosote-painted plank and tar paper. Dry wells and pipelines for roof drains shall be let under separate contract.

Article 203 – *Back Filling*

6. The contractor shall back fill against all walls and foundations and fill other places necessary to obtain lines as shown on drawings. All earth back filling shall be flushed with water and tamped every layer of 2 feet. After back fill has settled, contractor shall fill in all low spots.

Article 204 – *Finish Grading*

7. By separate contract.

SECTION 3 — MASONRY MATERIALS

Article 301 — *Delivery — Storage*

1. All materials shall be handled and stored so as to prevent damage by water or breakage. All packaged materials shall be kept in original containers until ready for use.

Article 302 — *Water*

2. All water used shall be free from injurious amounts of oil, acid alkali, organic matter, and other deleterious substances.

Article 303 — *Sand*

3. All sand shall be of a good grade of washed bank sand, sharp, well graded and free from salt, clay, and organic matter. It must be properly screened before mixing with cement or mortar.

Article 304 — *Portland Cement*

4. All Portland cement shall be of an approved brand and must stand the tests required by the A.S.T.M. designation C-150.

Article 305 — *Lime*

5. All lime shall be fresh hydrated lime of a standard product and shall contain not more than 5 percent ashes, clunker, or other foreign matter. All lime used shall comply with the standards of the A.S.T.M. specification C-6-49. All cement and lime shall be kept under cover and be raised from the ground on planks.

Article 306 — *Stone*

6. All crushed stone or gravel used shall be clean, hard crystalline rock, free from shale, clay or other soft material. Stone for plain concrete shall be well graded with maximum diameter of not over 2 inches. Crushed stone for reinforced concrete shall not exceed 1 1/2-inch diameter.

Article 307 — *Mortars*

7. All mortar measurements shall be by volume: sand and cement mixed dry, lime putty added, then enough water added to bring mixture to the proper consistency for use. No mortars shall be used after standing for more than one hour.

   - Cement Mortar — 1 part Portland cement, 3 parts sand, lime putty not over 25% of cement volume.
   - Cement Lime Mortar — 1 part Portland cement, 1 part lime putty, 6 parts sand.
   - Pointing Mortar for stone work — 1 part nonstaining cement, 2 parts white sand to which shall be added enough hydrated lime to make a workable consistency.
   - Fire-Brick Mortar — mortar for fire-brick shall be of fire clay.
   - Brick Mortar — mortar for all common brick shall be cement lime mortar, 1:1:6 mix. Mortar for face brick shall be cement lime mortar, 1:1:6 mix.

Article 308 — *Hearth Tile*

8. Hearth tile shall be 4″ x 4″ x 3/4″. It shall be hard-burned, but not vitrified and in no case shall it be soft or off-color. Color shall be selected by owner.

Article 309 — *Concrete Block*

9. All concrete block shall meet the standard specifications of A.S.T.M. C-90-44 and Federal Specifications SSC-621. These blocks shall be composed of refined Penn. anthracite cinders with a mixture of some local bituminous cinders; the balance of the mixture shall be a sharp clean sand and gravel with an appropriate amount of high early strength cement.

Article 310 — *Brick*

10. All common brick must be of good quality local, hard-burned brick of even color.

Article 311 — *Flue Linings*

11. Flue linings shall be made of fire clay and shall be hard-burned, free from workage, cracks, splits, chips, or other defects.

Article 312 — *Fire Brick*

12. All fire brick shall conform to the standard specification of the A.S.T.M. C-133.

SECTION 4 — CONCRETE WORK

Article 401 — *Proportions of Ingredients*

1. Concrete for all footings shall be 1 part Portland cement, 2 parts sand, 4 parts broken stone not larger than 2 inches in diameter. Concrete for all walls: 1 part Portland cement, 2 parts sand, 5 parts broken stone. Storage — garage and furnace room: two layers with lower layer 1:2:4 mix; top or finish layer, 1:2 mix. The floors in the workshop, basement, lavatory, and recreation room: top layer, 1:2; lower layer, 1:2:4 mix. All exterior slabs, floors, and aprons: 1:2:4 mix.

Article 402 — *Mixing Concrete*

2. All concrete shall be mixed by machine. Concrete shall be mixed until there is uniform distribution of all ingredients. The resulting concrete shall be of uniform color and appearance. The maximum time allowed between mixing and placing shall not exceed 30 minutes and no retempering will be allowed. All concrete shall be properly tamped, spading the stone away from all form work.

Article 403 — *Forms*

3. All necessary forms shall be provided, erected and braced by the contractor. All forms shall be of sound lumber, properly matched and wired and otherwise supported to insure tightness and rigidity. All loose blocks of wood to be removed from forms before pouring concrete. Forms shall not be disturbed until concrete has set sufficiently to carry its own weight and all other loads that may occur on the concrete.

4. If conditions are favorable, trench forms can be used instead of wood forms; however, sides must be clean, even, and vertical and bottoms must be true and level and free from fill.

5. Box out all openings for chases and slots used for wires, pipes, conduit, etc., and build into concrete all inserts, anchors, ties, and hangers as required.

Article 404 — *Curing*

6. All concrete in floors and slabs shall be kept damp continuously for one week after it has been poured.

Article 405 — *Surface Finish*

7. After removing forms, cut back all metal ties, wet and fill all voids and rough surfaces with cement mortar, 1:2 mix. Apply one coat of cement mortar, 1:3 mix, floated and trowelled using wood float.

Article 406 — *Tile Base*

8. The concrete base for tile floor in bathroom shall be composed of 1 part Portland cement, 2 parts sharp clean sand and 4 parts crushed stone. The crushed stone shall be well graded within the limits of 1/4 inch to 1/2 inch. This base shall be 1 inch thick with metal lath imbedded. Lath shall be metal reinforcement lath with triangular mesh of No. 8 wire on 4-inch centers, diagonally laced with No. 14 wire on 4-inch centers. Wood floor shall be completely covered with 15-pound pitch felt before covered with metal lath. The metal lath shall be fitted to entire floor area of bathroom, butted to side walls and lapped 2 inches at splices. Lath shall be fastened to wood floor with galvanized staples at 6-inch centers both ways.

Article 407 — *Gravel Fill*

9. Under all concrete floors and slabs, place bed of 1-inch diameter gravel; this gravel to be wetted and well tamped to the thickness as shown. No gravel bed to be less than 4 inches thick.

Article 408 — *Rubble Veneer*

10. All rubble walls as shown shall be Ohio-Brierhill natural sandstone in yellow and buff. All stones shall be well bonded, bedded and set down hard. Stones shall be accurately cut to fit where necessary and exposed flame shall be at right angles to the face where possible. Stone retaining walls shall be capped, using 5-inch stones, laid flat with 1-inch projection over wall. Joints shall be 1/2-inch. Veneer stones shall be as close to 5 inches thick as possible. Maximum height of stones, 12 inches; maximum length, 20 inches; minimum height, 4 inches; minimum length, 8 inches. All stones shall be cleaned and drenched before laying. All stone work shall at all times be adequately protected from damage. On veneer work, galvanized metal ties shall be used. These ties shall be spaced horizontally not over 2 feet apart, and vertically not over 18 inches. The contractor shall cover all wood surface walls with 15-pound pitch-saturated felt before stone veneer is applied. Felt shall be lapped at least 2 inches and secured to wood sheathing with 1/4-inch by 1/2-inch wood strips and galvanized nails. The face of all stone work shall be cleaned upon completion with an approved cleaning compound; no acid to be used on stone work.

SECTION 5 — BRICKWORK

Article 501 — *Chimneys*

1. The contractor shall build fireplaces and chimney to dimensions as shown with particular reference to 1/2-inch scale details. All flues throughout shall be lined with hard-burned terra cotta flue lining set in full mortar joints left smooth on inside with no projections or droppings. Furnace flue shall start not over 4–6 inches above cellar floor, size as shown. Widths shall be well filled and solid. All brick work shall be kept at least 1 1/2 inches from all floor framing and 3/4 inch from all studding. An approved type of terra cotta thimble of proper size shall be placed at proper location for furnace pipe. Furnish and install "Cover" ash trap in first floor fireplace. Each fireplace shall be provided with a "Donley" 242 fireplace damper of the proper size. The fireplace in the recreation room shall be faced with Belden Yorkshire Blend face brick, white mortar, flush joint. Hearth for this fireplace shall be 4-inch by 4-inch red tile, 1/2-inch joint, white mortar.

2. The living room fireplace shall be faced with HY-TEX Brick, Washington Colonials with racked out joints, using black mortar. The tile used for this fireplace hearth will be 4-inch by 4-inch, heathen brown quarry tile, 1/2-inch joint, using black mortar. The brick work on the chimney will be carried up to a point one foot below the roof. From this point to a height above the roof as shown on plans, the chimney will be topped with rubble all laid as per stone veneer specification. Cap chimney with cement mortar, 1:2 mix, carrying same to within 1 inch of top of flues. The contractor will build in at proper joints, 24-ounce copper flashing setting same a minimum of 1 1/2 inches into joint.

SECTION 6 — STONE FLAGGING

Article 601 — *Concrete Slab*

1. Flagging shall be laid on a well tamped porous fill at least 4 inches thick and a concrete slab 4 inches thick: concrete shall be 1:2:4 mix reinforced with 6-inch by 6-inch, #10 and #10 wire fabric. Slab shall have proper grade and pitch as shown on the drawings.

Article 602 — *Flagging*

2. When same has set, place 1-inch setting bed as required to bring flagging to true even finish with proper pitch as required. Flagging will be of 1-inch to 1 1/4-inch thick cut on site as required. Type of stone and colors will match stone walls and will be suitable to architect. Pointing mortar

shall be 1:2 mix, nonstaining. Pattern of flagging shall be semi-irregular, fitted, with all joints flush. All work shall be cleaned immediately after pointing, using fiber brushes and muriatic acid if necessary. Acid shall be thoroughly rinsed off with water.

SECTION 7 – DAMPPROOFING

Article 701 – *Dampproofing Walls*

1. The contractor shall furnish all labor and materials to complete all dampproofing as shown on the drawings or herein specified. The outside faces of all concrete walls below grade shall be given one coat of asphalt dampproofing. The dampproofing shall be brush applied. The material to be used is manufactured by Hydro Inc. for the purpose intended and is known as Hydro Tite #21.

2. The contractor shall examine all surfaces to be dampproofed and shall notify the architect in writing of any defects in same. Starting the work by the contractor shall imply acceptance of the condition as satisfactory.

3. Before applying the materials the contractor shall clean off all surfaces, leaving same clean and ready for dampproofing.

4. Material shall be applied as per manufacturers' directions. The number of coats specified is minimum and if all surfaces are not covered thereby, the contractor shall apply more material. Dampproofing shall not be applied in cold or wet weather or when the surfaces are damp. All concrete floors that are to be covered with asphalt tile shall be given a primer coat of Armstrong primer made by Armstrong Linoleum Company and applied as per manufacturers' directions.

5. All walls in the workshop, recreation room and basement lavatory shall be given one coat of Hydro Tite #21.

6. The walls of the furnace room and storage room shall be left unfinished.

SECTION 8 – MISCELLANEOUS METAL WORK

Article 801 – *General*

1. The contractor shall furnish all labor and material to complete all structural steel as shown on drawings or herein specified. All steel shall conform to the standard specifications of A.S.T.M. in the latest edition.

2. All steel lintels and girders as shown shall be set true and level. Girder shall have two bearing places 3/8-inch thick, 6 inches square. The contractor shall also place steel curb as shown and shall furnish and install steel hand rail as shown on rear porch.

Article 802 – *Roof Flashing*

3. All flashing shall be 26-gauge galvanized sheet metal. This shall include base flashing where roof abuts vertical surfaces; step flashing at pitched roof surface and vertical surfaces; valley flashing; flashing at pipes through roof; flashing at louvers and cupolas, gutters and leaders, all as shown on the drawings.

4. All workmanship shall be of the best trade practice.

SECTION 9 – TILE WORK

Article 1001 – *Preparatory Work*

1. The installation of concrete slabs, concrete fills, metal lath on bathroom floor and walls and similar supports for tile work are specified in other sections of these specifications.

2. The contractor shall furnish all labor and materials to complete ceramic tile, marble and quarry tile work as shown. All tile including setting beds shall include the following: fireplace hearths, bathroom floor, and bathroom walls. A marble threshold shall be furnished and set in bathroom.

Floor tile shall be ceramic 2-inch by 2-inch; color and pattern shall be chosen by owner. Wall tile shall be glazed 4-inch by 4-inch. In tub stall, wall tile will be carried to a height of 6 feet with molded cap. Other bathroom wainscoting shall be four feet high, with base plinth blocks and molded cap. The contractor will furnish and set one towel bar, one soap holder, one tissue insert, and one assist bar, all of approved shape and color. All work under this division shall be thoroughly cleaned at completion and protected from damage.

SECTION 10 – CARPENTRY

Article 1001 – *Framing Materials*

1. All framing material herein referred to shall conform to American Lumber Standards as specified in Simplified Practice Recommendation R16-29 Lumber (4th edition) issued by the Bureau of Standards of the U.S. Department of Commerce, Washington, D.C. It shall be dressed on four sides to standard dimensions and shall be free from defects or imperfection that might impair its durability or strength. The contractor must furnish and assemble any timber of the proper size and kind necessary to the completion of the building whether specifically mentioned or not. All work shall be framed in a thorough and quality manner. All sills, if solid, shall be halved and lapped at all corners. Trim all wells and openings and in no case shall headers and trimmers be less than two inches from a chimney. All headers, trimmers, and studs around openings must be durable. All framing lumber shall be equal to No. 1 common grade.

Article 1002 – *Sizes*

2. The general sizes of rough lumber shall be as follows:
    Sills 2″ x 4″, 4″ x 6″ FIR
    Girders 6″ x 8″, 8″ x 10″ FIR
    Studding 2″ x 4″ FIR
    Joists 2″ x 8″, 2″ x 10″ FIR

3. Common Rafters, 2″ x 6″; Hip-Valley Rafters 2″ x 8″ FIR
    Plates 2″ x 4″, Doubled 2″ x 6″ FIR
    Bridging 2″ x 3″ FIR
    Furring 1″ x 2″ SPRUCE
    Sheathing 1″ x 6″ N.C. PINE
    Firestops 2″ x 3″, 2″ x 10″ FIR

Article 1003 – *Cutting*

4. The contractor shall do all cutting of woodworking to accommodate the work of the other trades but no cutting shall be done that will impair or weaken the structure.

Article 1004 – *Sills*

5. Sills shall be set level and true and shall set on a bed of cement, 1:3 mix. Sills shall be anchored with 5/8-inch by 18-inch anchor bolts, spaced not over 6 feet on centers. Anchors shall also be set 18 inches from each corner.

Article 1005 – *Girders*

6. Girders shall be set true and straight with 1/2-inch crown on 20 feet. All girders shall be flush type with 2-inch by 3-inch cleat, securely spiked.

Article 1006 – *Plates*

7. Wall and bearing partition plates shall be double; nonbearing partition plates, single. The cap plates on concrete block walls in basement shall be 4-inch by 6-inch single, anchored to wall with 1/2-inch by 12-inch bolts, spaced not over 4 feet o.c.

Article 1007 – *Joists*

8. Joists resting on wood sills shall rest directly on same. Joists resting on or against steel I beams or flush type girts or girders shall be notched to proper depth.

9. Joists shall be set with crowning side up. Double all joists parallel to and under partitions. Such doubled joists shall be separated with 2-inch blocking. Space joists under bathroom floor 12 inches OC.

Article 1008 — *Joist Cross Bridging*

10. Cross bridging shall be made of 2-inch by 3-inch strips with ends beveled and double crossed. Spacing shall be according to local building code, or, 1 row on spans up to 10 feet, 2 rows on spans 10 feet to 18 feet. Nail each end of bridging with two 8d common nails. Bottom ends shall be nailed after subfloor is in place.

Article 1009 — *Studs*

11. Arrange three studs at all corners. Double all studs at sides of openings, inside stud to support header. All studs shall be spaced 16 inches o.c. unless noted otherwise and unless other spacing is required for ducts and pipes. Frame all stud openings to proper size to secure door and window frames and jambs.

Article 1010 — *Stair Carriages*

12. Basement stair stringers shall be supported with vertical 2-inch by 4-inch studs extending to basement floor at mid-distant of the stringer length. Attic stair stringers shall be supported with nails into partition studs, mid-distant of the stringer length.

Article 1011 — *Rafters*

13. Rafters shall be spiked to place and cut accurately to detail to secure a solid bearing. In unfinished attic where no ceiling beams occur, provide collar beams at 3 feet above attic floor, 1-inch by 6-inch, well nailed at every second rafter. Cut all rafters to form all pitches as shown.

Article 1012 — *Firestopping*

14. Firestops shall be at least 2 inches thick and shall be arranged so as to cut off all concealed draft openings and form an effectual draft barrier between stories and roof spaces.

Article 1013 — *Shingling*

15. All shingles to be asphalt, strip type, 200# Flintkote or equal laid as per manufacturer's directions, using large head galvanized nails. Shingles shall be cut accurately at valleys, intersections, ridges, and hips. Starter course shall consist of 1 row 16-inch red cedar shingles and 2 rows of asphalt shingles.

Article 1014 — *Exterior Finished Carpentry*

16. The contractor shall furnish and install all exterior finished carpentry shown on the drawings or specified herein or both as follows: all lumber shall be white pine, C select, unless otherwise specified. All moulding shall be stock patterns, sizes as shown.

Article 1015 — *Cornice*

17. The contractor shall build the cornice according to details, using 2-inch by 3-inch lookout blocks at every rafter. Plancier to be 3/8-inch composition board.

Article 1016 — *Vertical Siding*

18. Siding shall be T X G bevel edge, random widths accurately scribed and fitted. All edges and ends in contact with stone work flashing shall be given one coat of white lead.

Article 1017 — *Clapboards*

19. Clapboards shall be 8-inch Western Red Cedar, neatly fitted against all corner boards and frames. Clapboard courses to be not in excess of 6 inches of exposure.

Article 1018 — *Cased Post*

20. As shown, shall be composed of two 2 by 4's spiked together and cased with 3/4-inch by 6-inch white pine.

Article 1019 — *Water Table*

21. Install 1 1/8-inch by 6-inch water table where needed to form first course for clapboards. No drip cap to be used but top edge of water table shall be beveled 1/8-inch.

Article 1020 — *Trellises*

22. The contractor shall furnish and install all trellis work as shown according to details. Fasten all trellis work firmly to concrete and stone work with 20d spikes driven into lead sleeves.

Article 1021 — *Cupola*

23. The cupola shall be completely assembled at the mill as per details with roof and walls of 3/4-inch waterproof fir plywood, exterior grade. The entire cupola shall be given a primary coat of white lead and oil before being placed in position. Install metal louvers. The cupola shall be accurately fitted to roof and neatly flashed with 26-gauge galvanized sheet metal. Cupola roof shall be 16-ounce copper.

Article 1022 — *Louvers*

24. The contractor shall furnish and install louvers as per details. Louvers to be mill assembled and given a coat of white lead paste before installation. The louvers shall be accurately fitted to all roof planes and shall be flashed with 26-gauge galvanized sheet metal to make a watertight job. The copper roof on the round top louver shall be let under a separate contract.

Article 1023 — *Windows — Wood*

25. All double-hung frames and sash shall be Andersen or Morgan, clear white pine, except pulley stiles and parting beads which will be of yellow pine. Sash to be 1 3/8 inches thick and fitted with tubular type spring balancers. All sash sizes as per window schedule. All wood casement frames and sash to be "Curtis Insulated Silentite" with stock for type of window specified.

26. All basement window frames and sash shall be Andersen or Morgan pine with movable sash.

Article 1024 — *Exterior Door Frames*

27. Door frames, except garage, shall be Curtis or Morgan of white pine, rabbeted for 1 3/4-inch doors. Door jamb reveal shall not be less than 1 1/8 inches. Door sills shall be of oak. Garage door frames shall have 3/4-inch W.P. jambs, installed as per directions given by manufacturer of overhead doors.

Article 1025 — *Ceiling Boards*

28. The ceiling boards over porch and front entrance shall be of 1-inch by 6-inch T. & G. white pine.

29. All exterior finished carpentry shall be erected plumb, level, true and in accord with all details. All members shall be erected in as long pieces as is practical with joints arranged to be as inconspicuous as possible. All work shall be well nailed with all nails countersunk. All work to be left free from defects and in good condition to receive the finish.

Article 1026 — *Interior Finished Carpentry*

30. The contractor shall furnish and install complete, all interior finished carpentry work, as shown on the drawings or herein specified, or both. All woodwork, trim, cabinets, etc. shall be of clear Idaho white pine, "B" grade, unless otherwise specified. Lumber shall be protected from damage, both in transit and at the job site. Material shall not be delivered unduly long before needed for the proper conduct of the work. Lumber shall not be stored or installed in the building until ten days after the completion of plastering.

Article 1027 — *Jambs*

31. Jambs shall be full width of partitions, with all edges back beveled, and sides rabbeted to receive heads. Set minimum of five shim blocks per side. Blocks shall be placed behind door hinge locations.

Article 1028 — *Casings*

32. All door and window casings shall be 2 1/2-inch face, pattern to be suitable to owner. Trim shall be assembled at the mill with corners mitered and splined. Casings shall be erected plumb, square and true with 1/4-inch reveal on door frames.

Article 1029 — *Doors*

33. All door sizes shall follow door schedule: all exterior doors, 1 3/4 inches thick; all interior doors, 1 3/8 inches thick except as herein noted. All exterior doors shall be #1 pine "Curtis" or Morgan of patterns approved by the architect.

Article 1030 — *Doors — Interior*

34. All interior flush doors shall be of an approved manufacture, hollow core, of white pine cross banded with 1/8-inch veneers and faced vertically on both sides with 1/20-inch thick veneers of white pine. Sliding doors shall be seven ply white pine plywood, complete with hangers, guides and track of an approved type. Batten doors shall be five ply white pine plywood, backed with three 6-inch horizontal battens.

Article 1031 — *Doors — Garage*

35. Garage doors shall be of overhead type, roll up, four section, of approved manufacture, properly fitted and hung and controlled by overhead mechanism of approved type. All garage door hardware shall be fitted and placed as per manufacturer's directions.

Article 1032 — *Base*

36. Install 1-inch by 4-inch base, mitering all external and coping internal corners. Shoe mold, 1/2-inch ogee pattern; base mold, C 5451. The base mold shall be continuous around radiation openings. Base in closets 1-inch by 4-inch with no base molding.

Article 1033 — *Mantels*

37. Furnish and install special mantel on Living Room fireplace as per detail, tapered surround of white pine.

Article 1034 — *Ironing Board*

38. Furnish and install one R 9204 or similar built-in ironing board with flush plywood doors.

Article 1035 — *Closets*

39. All closets will be furnished with 1-inch by 12-inch shelves, maximum amount of 1-inch by 3-inch hook strip and one closet pole of fir, 1 1/2-inch diameter. All linen closets shall be fitted with shelving of maximum widths spaced not over 12 inches.

Article 1036 — *Medicine Cabinet*

40. The medicine cabinet shall be of steel, made by Baxter Manufacturing Company, #GR 901.

Article 1037 — *Kitchen Cabinets*

41. Kitchen cabinets and cupboards shall be mill assembled, sizes and lengths as per details. All rails and stiles 2-inch. All exposed ends of 1/2-inch plywood. Doors and drawer fronts of 3/4-inch plywood, lipped 3/8-inch. Drawer sides and backs, 1/2-inch white wood; drawer bottoms, 1/4-inch fir plywood. Provide all lower cabinets with 3-inch by 3-inch toe space. Counters shall be 3/4-inch waterproof fir plywood, 36 inches high, covered with Formica or equal as approved. All cabinet doors to be fitted with magnetic catches. Prefabricated Morgan cabinets or equal may be used in place of above.

Article 1038 — *Paneling*

42. The walls of the lavatory and workshop shall be covered with 3/8-inch fir plywood. Place 1-inch by 3-inch chair rail, 4 feet above floor in lavatory and cover all plywood joints with 1/4-inch by 1-inch plain white pine panel strips. Run 3-inch crown mold at ceiling. All vertical plywood joints to be accurately fitted. Ceiling molding, 3-inch cove.

43. The paneling in the recreation room and hall leading to lavatory shall be knotted pine random widths, in continual lengths from floor to ceiling. Knotty pine shall be tongue and grooved, matched with V joints, and shall be blind-nailed.

Article 1039 — *Stairs*

44. The contractor shall furnish all labor and materials to complete all wood stair work as shown on the drawings and herein specified, or both. All material shall be stock of an approved quality. All stair work shall be cabinet work finished and installed by skilled mechanics. Cellar stair treads of oak with W.P. risers and skirt boards. Attic stair treads of oak with W.P. risers and skirt boards. All treads and risers shall be tongued and grooved and skirt boards or stringers shall be housed. All work to be wedged and glued. Both stairs shall have one 1 3/4-inch round stair rail each complete with two rosettes and three rail brackets. All work to be hand sanded and left ready for painter.

Article 1040 — *Finish Floors*

45. The contractor shall furnish and lay wood finish floors in the living room, bedrooms, dining area and connecting halls. All flooring shall be #1 clear white oak, 2 1/4-inch face. All flooring shall be kiln dried and shall not be brought to the job until plaster is dry. Under all finish floor place a layer of 15# asphalt-saturated wool roll felt and lap all joints 2 inches. All wood floors to be machine-sanded with all edges hand-scraped and hand-sanded, and left in good condition for finishing.

SECTION 11 — PAINTING AND FINISHING

Article 1101 — *Delivery and Storage*

1. Delivery — Painting, varnishing, and finishing materials shall be delivered to the site in unopened, original containers, bearing manufacturer's printed labels.

2. Varnish — All varnish shall be delivered to the job in the manufacturer's original containers, bearing labels denoting the quality, brand and trade name. Containers of over five-gallon capacity shall not be used. Said containers shall be bulked in convenient size cartons for delivery. The cartons shall be effectively sealed by the manufacturer of the varnish and the seal shall be unbroken when the varnish is delivered to the job.

3. Storage — The materials shall be stored at the site where directed. The storage space shall be kept clean. Oily rags shall be burned or removed from the premises at the close of each day's work and all other necessary precautions taken to avoid damage by fire.

Article 1102 — *Materials*

4. Standard brands of the highest quality shall be selected by the owner with the approval and guidance of the architect.

5. An approved Painting and Finishing Schedule shall be set forth on the drawings to assist the contractor to estimate costs. This schedule shall show types of finishes on all surfaces to be painted, room and space identification.

Article 1103 — *Preparation of Surfaces*

6. Bare metal shall be thoroughly cleaned of all foreign matter, rust, and dirt before paint is applied.

7. Shop painted metal shall be cleaned and the shop coat retouched where marred, before succeeding coats are applied. Oil or grease shall be removed with a suitable solvent.

8. All painting on hollow metal shall be thoroughly sandpapered and any dents or imperfections shall be filled with knifed or metallic filler.

9. Woodwork to be painted or varnished shall be in perfect condition before being coated. The surface shall be clean, smooth and dry.

10. Before acid stain is applied to open grain hardwood, the surfaces shall be lightly sponged with cold water. After the wood has dried, the raised grain shall be removed and the surface smoothly sandpapered. The stain shall be freely applied with a soft-hair brush, repeating the brushing, or wiping so as to remove any blotching and uneven surfaces, producing a uniform color.

11. All scratches, dirt, stains, raised grain, or other surface defects, shall be removed before the painting or varnishing is started.

12. Painting or varnishing shall not be done under conditions of weather or temperature unsuitable for executing a first-class job. The atmosphere must be free from dust and dirt, preventing the lodgment of foreign matter in the fresh paint or varnish. Floors must be broom clean before painting is started.

13. Woodwork to be painted shall have all knots, pitch streaks and bad sap spots shellacked two coats before priming coat is applied.

14. When surfaces to be painted or varnished are defective or unsuitable for coating, the contractor shall prepare such surfaces as directed by the architect.

Article 1104 — *Workmanship*

15. Painting — Priming coats on woodwork, except for woodwork where priming is required at the shop, shall be applied as soon as possible after the work is in place. All coats of paint shall be laid on so as not to show brush marks. The top and bottom edges of all wood doors which extend down to the floor shall be given two coats of paint. All other edges of doors shall be finished the same as face of doors. The top and bottom edges of all wood sash shall be given two coats except that top of meeting rails shall be finished the same as interior face of sash. Shop coat on metal shall be applied by brush or spray.

16. No paint shall be applied on sliding contacts or similar surfaces where bare metal is necessary for the proper operation of the unit and any paint applied to such surfaces shall be removed as a part of the work under this section of the specifications.

17. Undercoats on interior wood or metal finishes shall be thoroughly sanded with Number 00 Sandpaper, or other equal abrasive, removing all surface defects and providing a smooth, even surface for subsequent coats.

18. Varnish and Enamel shall be evenly and smoothly flowed on, and shall show no sags or runs.

19. Varnish shall be used as it comes from the manufacturer's container, without thinning or adulterating. The varnish shall be flowed on, and all coats except the last shall be sandpapered.

20. Staining — Surfaces specified to be stained shall be covered with a uniform application of stain, equalized where necessary and wiped off if required.

21. Shellac shall be evenly applied producing a uniform coating.

22. Putty Stopping — Nail holes, imperfections and defacement in both exterior and interior wood finishes shall be putty stopped after priming coat of paint, filler or shellac has been applied. Putty stopping shall be brought flush with the finished surface in a neat and workmanlike manner.

23. Protection — Work subject to damage or defacement shall be properly protected. Floors, stair treads, landings and the like shall be covered with heavy building paper or cloth properly secured in place. At acceptance, the paint and varnish finish work must be in a neat, sound, undamaged condition.

24. Cleaning — Upon completion of the paint and varnish work, all surplus materials, empty packages, debris and the like shall be removed from the building and premises. All finished work shall be retouched where necessary, thoroughly cleaned and left in a neat, perfect condition. Daubs or spatterings of paint or varnish, shall be removed from hardware, glass, floors, walls or any other surface thus disfigured.

SECTION 12 – ROUGH HARDWARE

Article 1201 – *Rough Hardware*

1. The contractor shall furnish all nails, spikes, bolts, screws, hangers, stirrups, anchors, ties and other accessories shown on the drawings or as required to secure the woodwork properly and all such accessories shall be of design required to develop the full strength of the members to which they are attached. Anchors or other accessories required to be built in with the masonry shall be furnished in ample time and the setting of same given all necessary attention to insure their proper location.

2. Copper nails shall be used for all exterior nailing on cupolas, louvers, and trellises unless the woodwork is completely covered with copper.

SECTION 13 – FINISH HARDWARE

Article 1301 – *Materials*

1. The owner shall select the type, color and material of all finish hardware with the guidance of the architect who, in turn, shall set up a hardware schedule showing the locations of installation and keying of door locks and all other required hardware. Schedule of hardware shall be printed on the drawings for the convenience of the contractor in estimating and clear understanding of locations.

2. Contractor shall install all finish hardware.

INDEX

The seven foldouts attached contain the following:

House plans for the assignments:

House plans for the Comprehensive Review: